Agrarian Systems and Climate Change

Agrarian Systems and Climate Change

Journeys of adaptation in the Global South

_____

Edited by

Hubert Cochet, Olivier Ducourtieux, and Nadège Garambois

CABI

CABI is a trading name of CAB International

CABI
Nosworthy Way
Wallingford
Oxfordshire OX10 8DE
UK

CABI
200 Portland Street
Boston
MA 02114
USA

Tel: +44 (0)1491 832111
E-mail: info@cabi.org
Website: www.cabi.org

Tel: +1 (617)682-9015
E-mail: cabi-nao@cabi.org

Originally published in French under the title Systemes Agraires et Changement Climatique au Sud by Hubert Cochet. © Éditions Quæ, 2019.

A catalogue record for this book is available from the British Library, London, UK.

The views expressed in this publication are those of the author(s) and do not necessarily represent those of, and should not be attributed to, CAB International (CABI). Any images, figures, and tables not otherwise attributed are the author(s)' own. References to internet websites (URLs) were accurate at the time of writing.

CAB International and, where different, the copyright owner shall not be liable for technical or other errors or omissions contained herein. The information is supplied without obligation and on the understanding that any person who acts upon it, or otherwise changes their position in reliance thereon, does so entirely at their own risk. Information supplied is neither intended nor implied to be a substitute for professional advice. The reader/user accepts all risks and responsibility for losses, damages, costs and other consequences resulting directly or indirectly from using this information.

CABI's Terms and Conditions, including its full disclaimer, may be found at https://www.cabidigitallibrary.org/terms-and-conditions.

A catalogue record for this book is available from the British Library, London, UK.

ISBN-13: 9781800628113 (hardback)
9781800628120 (OA ePDF)
9781800628137 (OA ePub)

DOI: 10.1079/9781800628137.0000

Commissioning Editor: David Hemming
Editorial Assistant: Helen Elliott
Production Editor: James Bishop
Translator: Kevanne Monkhouse

Typeset by Straive, Pondicherry, India
Printed and bound in the USA by Integrated Books International, Dulles, Virginia

Contents

_____

Contributors

_____

Elsa Champeaux
Expert, Ginger SOFRECO, Clichy, France
elsa.champeaux@gmail.com

Hubert Cochet
Professor of Comparative Agriculture, Comparative Agriculture Research & Training Unit, UMR Prodig, AgroParisTech, 22 place de l'Agronomie, 9110 Palaiseau, France
hubert.cochet@agroparistech.fr

Olivier Ducourtieux
Associate Professor of Comparative Agriculture, Comparative Agriculture Research & Training Unit, UMR Prodig, AgroParisTech, 22 place de l'Agronomie, 9110 Palaiseau, France
olivier.ducourtieux@agroparistech.fr

Samir El Ouaamari
Associate Professor of Comparative Agriculture, Comparative Agriculture Research & Training Unit, UMR Prodig, AgroParisTech, 22 place de l'Agronomie, 9110 Palaiseau, France
samir.elouaamari@agroparistech.fr

Mathilde Fert
Agro-economist, 2306 route de Marcerin, 64300 Argagnon
fert.mathilde@gmail.com

Nadège Garambois
Associate Professor of Comparative Agriculture, Comparative Agriculture Research & Training Unit, UMR Prodig, AgroParisTech, 22 place de l'Agronomie, 9110 Palaiseau, France
nadege.garambois@agroparistech.fr

Thérèse Hartog
Program Officer, ODEADOM, 3 bis villa Jean Godard, 75012 Paris, France
therese.hartog@odeadom.fr

Thibault Labetoulle
Agro-economist, 1 route la Fontaine, 86140 Scorbe Clairvaux, France

Esther Laske
Program Officer, The World Bank, Washington D.C., USA
esther.laske@gmail.com

Philippe Le Clerc
Rural development and Food security Program Manager, European Commission, Delegation of the
    European Union to Mauritania
philippe_le_clerc@hotmail.com
Ulysse Le Goff
Researcher, FiBL Lausanne, Avenue des Jordils 3, CH-1006 Lausanne, Switzerland
ulysse.legoff@fibl.org
Jean-Luc Paul
Associate Professor of Anthropology, Institute of African Worlds, University of the French
    Antilles, France
jean-luc.paul@univ-antilles.fr
Léa Radzik
Study Officer, APCA Chambers of Agriculture France, 8 rue des Iris, 75013 Paris, France
lea.radzik@gmail.com
Céline Tewa
Founder of the Zero Waste 'Les Gamelles' network
celine.tewa@lesgamelles.fr
Lucie Thibaudeau
Agriculture and Rural Development Project Officer, French Development Agency (AFD),
    Rabat, Maroc
thibaudeaul@afd.fr
Louis Thomazo
Agro-economist, Via di Santa Chiara, 42, 00186 Roma, Italy
louis.thomazo@studiumversailles.fr
Charlotte Verger-Lécuyer
Advisor, Cerfrance49, Saumur, France
charlotte.verger1@gmail.com
Niel Verhoog
Project Manager for Organic Agriculture and Agricultural Vocational Training, CEZ-Bergerie
    Nationale de Rambouillet, Parc du chateau, CS 40609, 78514 Rambouillet cedex
niel.verhoog@educagri.fr
Florie-Anne Wiel
Project Manager, FADEAR, Bagnolet, France
florieanne.wiel@gmail.com

Introduction

Hubert Cochet*, Olivier Ducourtieux, and Nadège Garambois
Comparative Agriculture Research & Training Unit, UMR Prodig, AgroParisTech, Palaiseau, France

The studies and publications on the risks associated with climate change to agriculture are extremely numerous. A quick review of this abundant literature makes it possible to identify the natural hazards and risks linked to climate change that most affect—in frequency and intensity—the practices and production of farmers around the world. There is drought, an isolated hazard; aridification, a condition worsened by the hazard's increasingly frequent repetition until it becomes permanent; flooding, an isolated hazard; submergence, particularly through marine transgression; rising temperatures, which affect the crop and livestock calendar, or even their latitudinal geographic distribution (IPCC, 2014b), as well as yields (IPCC, 2014a), the ways harvests are preserved, and biological balances. Moreover, the effects of climate change manifest at multiple spatial scales, from region to cell via plant (or animal), field, ecosystem, and territory. They also affect economic exchanges, health (Springmann *et al.*, 2016), and social relationships. Climate change, with its diversity of phenomena and multiscale and diachronic dimensions, has systemic effects on agriculture (IPCC, 2014a,b; Altieri, 2016).

IPCC experts also emphasize, particularly in *Food Security and Food Production Systems* (Chapter 7 of the second volume), farmers'

adaptive capacity, especially in developing countries, and their past experience accounting for climate change in their farming practices, as well as the need to consider local knowledge in the search for adaptive solutions. The experts insist on the socioeconomic dimension of vulnerability and the need to go beyond purely technological approaches to adaptation in favor of approaches based on building resilience (IPCC, 2014a). For a given "global" change whose impact is perceptible on the regional or country scale, its consequences, along with productive systems' vulnerability or resilience and actors' capacity to adapt, differ according to the technical, social, and economic conditions in which they are found. They depend to a large extent on such conditions as farmers' access to resources (land, water), means of production, knowledge, markets, and the ways in which production factors are combined at the farm level and, on a larger scale, the agroecosystem level. This results in impact/adaptation trajectories that are themselves different.

This book is devoted to this link between global change and impacts and adaptation at the local level, combining a systemic approach and connecting different scales of analysis. We will attempt to unravel the complex ties between the scenarios developed at very large scales of analysis

*Corresponding author: hubert.cochet@agroparistech.fr

©2024 CAB International. *Agrarian Systems and Climate Change: Journeys of adaptation in the Global South* (eds H. Cochet, O. Ducourtieux, and N. Garambois)
DOI: 10.1079/9781800628137.intr

(global, continental, regional, etc.) and farmers' concrete experiences, lived at the territorial level. In addition to this great discrepancy in terms of scales of analysis and understanding of processes, there is a need to relate the multi-generational scale of possible climatic changes to that of agricultural practices carried out on the agricultural season level.

Studying climate change's possible consequences on agrarian systems and farmers' modalities and conditions of adaptation also requires a resolutely transdisciplinary approach. In this regard, Charlotte Da Cunha and Vanderlinden (2014) emphasize the need to recognize the complexity of social-ecological systems (SES) in order to address adaptation to global change and point to the importance of transdisciplinarity in addressing this issue: "it is thus necessary to build a space for integrating natural sciences, social sciences and humanities in the analysis of the dynamics of adaptation to global change" (p. 279). By mobilizing tools and concepts from both life sciences (ecology, agronomy, zootechnics) and social sciences (geography, economics, anthropology, history) around a subject (the transformation of agriculture in relation to global changes), we will attempt to use our comparative agriculture approach to understand the dynamics of farmers' adaptation to climate change.

This book does not, of course, aim to cover all the ground on such a complex issue. Through its methodological approach, which is as close as possible to "the field," and by relying on a detailed study of past and current agrarian dynamics in several regions of the world, it will—we hope—be able to contribute to the ongoing debates on this question.

I.1 Vulnerability, Resilience, and Adaptation

I.1.1 Vulnerability: a social science concept

The concept of vulnerability emerged in the social sciences and today is still "restricted" to this (broad) field. It has multiple definitions across disciplines, from sociology to political ecology to history and geography (Becerra, 2012; Buchheit

et al., 2016). Vulnerability first appears as the propensity to suffer from changes but can also be defined as lacking the right to access resources and the inability to mobilize them to avoid suffering a hazard's deleterious effects (Adger *et al.*, 2003; Janin, 2006; Becerra and Peltier, 2009; Brown, 2016; Buchheit *et al.*, 2016). Vulnerability can then be defined (measured?) as the product of the degree of exposure to hazards via the sensitivity to those hazards and the ability to cope with them—also known as responsiveness (Becerra and Peltier, 2009).

With respect to climate change, many studies and research focus primarily on the physical dimension of hazards, improperly named "natural risks" (IPCC, 2014a,b). The study of vulnerability allows us to put these technical-scientific approaches focused on a hazard's causal factors into perspective, in order to consider the social constructs that make individuals and societies more or less exposed to this hazard and adapt to or prevent the resulting risks (Becerra, 2012; Barnes *et al.*, 2013). This implies an interest in not only individuals' reactivity but also the social relations that condition access to resources and different social groups' capacity to react (Lallau and Thibaut, 2009). A "natural risk" is certainly a physical hazard, but its magnitude is based on the society's robustness and thus the history of the social relations that shaped it (Becerra and Peltier, 2009; Bonneuil and Fressoz, 2016).

The vulnerability of one group or another is not a datum in itself but rather the result of differentiation processes and impoverishment trajectories whose evolution and root causes must be understood. Indeed, climate change is also a kind of indicator of development inequalities in a given society. We will thus see, from the in-depth case studies presented in this book, that climate change is not in itself responsible for poverty. Studying its contribution to the fragility of societies, or of certain groups therein, requires a holistic approach that seeks to factor in together different global changes, their impacts, the greater or lesser vulnerability posed by some changes over others, and the capacities to adapt to these changes. Trying to isolate one of these factors of change (i.e. climate change) is therefore hardly conceivable, as reasoning from an "all else being equal" perspective presents too many oversimplifications. As we will see in Chapter 8, this type of oversimplification nevertheless inspires many public policies on adaptation.

I.1.2 Resilience: from materials to societies via ecosystems

The scientific use of the term resilience is old, initially employed in engineering and materials science to describe shock resistance. The term quickly went beyond this physical framework to be used in thermodynamics, psychology, computer science, art, and so on. In ecology, it describes an ecosystem's robustness, which is its capacity to recover its previous state after a disturbance more or less quickly and more or less completely. In this respect, the concept of resilience is at the heart of the approaches developed in the UN Millennium Ecosystem Assessment. In addition to ecology, the notion is found in risk science (cindynics) and, again by analogy, the social sciences: political science (including political ecology), economics, sociology, geography, and so on (Mathevet and Bousquet, 2014; Reghezza-Zitt and Rufat, 2015; Brown, 2016, Buchheit *et al.*, 2016). We see that resilience, even more than vulnerability, is a polysemous, polyconceptual term. This richness explains its success—that is when it does not cause confusion.

Resilience can describe a property, "an intrinsic quality of a system, an ability that shows itself at the moment of the impact," a process, "the succession of consecutive responses to the disturbance," or a result, "the system is able to rebound and deal with the threat" (Reghezza-Zitt and Rufat, 2015). Resilience is thus a systemic characteristic at different levels: ecosystem, society, social group, household, and individual (Altieri, 2016; Brown, 2016). The analogy with a material's elasticity is also used to define an ecological or social system's limits of resilience when a hazard exceeds the absorptive capacity of the system's weak point (Lallau and Thibaut, 2009). A levee provides effective protection from flooding, provided that the water height does not exceed that of the levee's lowest point.

Strictly dependent on access to resources and social differentiation, a society's resilience and vulnerability to climate change can only be studied, understood, and assessed by taking very fine-grained account of diversity at the local scale (IPCC, 2014b; Tasser *et al.*, 2017), both environmental (ecosystems and their stories, Locatelli, 2016) and socioeconomic (Brémond *et al.*, 2009; Hiwasaki *et al.*, 2014; Boit *et al.*, 2016). Considered in the long term, resilience goes beyond mere shock elasticity and the return to a previously hazard-resilient state (Bidou and Droy, 2009; Lallau and Rousseau, 2009; Chew and Sarabia, 2016; Brown, 2016). These transformations are long term, even irreversible, and highlight adaptation to a degrading climate (Mullenite, 2017).

I.1.3 From resilience to adaptation

Adaptation to climate change is the expression of individuals' and social groups' (in this case, farmers') ability to modify their practices in a changing environment to maintain (or increase) their standard of living and security despite increasing risks (IPCC, 2014b; Mullenite, 2017). These changes take time (Chew and Sarabia, 2016) and are not just about maintaining a metastable state: adaptation, in the long run, differs from resilience and vulnerability; adaptive capacities are not intrinsically identical to the conditions for resilience.

In this book, we are interested in the effects of climate change on farmers and their production and living conditions. When it comes to "adaptation," we must avoid hasty conclusions and exaggerated personifications of the term. Economic agents—farmers in particular—seek to adapt to climate change in order to avoid its deleterious effects on their living conditions, modifying their practices and environment within the limits of their means and capacities. On the other hand, an ecosystem or territory does not adapt; it is the various social groups interacting within them that have—or do not have—an active capacity to adapt.

Adaptation is an incremental and multiscalar process (Cash *et al.*, 2006; Chew and Sarabia, 2016), from changing varieties or sowing dates to the agricultural revolution in an agrarian system: adaptation is a systemic change (Altieri, 2016; Brown, 2016; Buchheit et al., 2016; Chew and Sarabia, 2016). Moreover, farmers' adaptation to climate change also depends on public action, including policies to reduce their vulnerability and increase their capacity (La Branche, 2011; Dupuis, 2015; Huang and Lee, 2016). Adaptation, like vulnerability and resilience, is largely based on social relations, power, and exchange in a given society, all for which the empowerment of actors and modes of governance—including community-based resource management—are essential (Cash et al., 2006; Poteete et al., 2010).

Within vulnerability research, many works have shown how and why the poorest social groups were the most vulnerable to climate change. However, little has been done to study how these same social groups adapt and what factors and mechanisms are at play to increase their resilience (Leichenko and Silva, 2014).

This indeed seems to be the neglected step-child of adaptation research. Research priorities are clearly elsewhere, as evidenced by the report by Rhodes and colleagues (2014) addressing key challenges and gaps in climate change adaptation research and policy in the West African agricultural sector. Far from placing the least importance on local knowledge, Rhodes and colleagues write, for example, "people's knowledge is limited or inadequate." Perhaps most appropriately, they continue, "There are few or no reliable methods for predicting the onset of the rainy season and intra-seasonal variability, nor are there methods for making weather forecasts as useful as possible for smallholders" (Rhodes et al., 2014, p. 40). Yet, when it comes to research "priorities," this report makes no mention whatsoever of other research areas, particularly in the social sciences and those dealing with the processes of differentiation, inequality, and vulnerability.

I.2 From Global Questioning to the Lessons of a Local and Contextualized Approach

I.2.1 Favoring a local approach

Public policy documents addressing the issue of climate change impact and adaptation often refer implicitly to scientific results which, based on the processing of vast databases (climatic, statistical, etc.), attempt to establish, with varying degrees of success, correlations between climate change and changes in yields, poverty rates or GDP points. In our case, rather than relying on statistics based on administrative units that are too broad to reflect the diversity of reality and whose data may be unconfirmed, we have chosen to work at the local level, that of the small agricultural region. This is the only scale that allows for in-depth immersion and meticulous study of agricultural and livestock practices, as well as the individuals and social groups that develop them. Observation of the landscape and practices, and dialogue with those who implement them, have thus inspired all the fieldwork whose results are presented in this book.

I.2.2 The interlocking of scales and the systemic approach

Careful observation of practices is carried out in the field, corral, pasture, water of the rice fields, privacy of the farmyard, and so on. Understanding these practices requires long discussions with those who put them into action, and sometimes demonstrations or re-enactments when direct observation is not possible. As for interpreting them, this calls for recourse to the two concepts of cropping system and livestock system. The first is relevant on a field scale, the second is on that of domestic animal herds. For example, we will see how in certain cases, climate risk mitigation strategies cannot be assessed without a detailed knowledge of intra-field microheterogeneities and the practices implemented to take advantage of them.

The analysis of a cultivated area in cropping system terms, or a domestic animal herd in livestock system terms, integrates many elements encountered at the production unit level: equipment and labor force, for example. This is why a detailed analysis of practices cannot be conducted without reference to a higher and more encompassing level of analysis, that of the production system.[1] It is at this level of analysis that one can understand how access to different ecosystems, the combination of different cropping systems, and, where applicable, different livestock systems, make it possible to "not put all one's eggs in the same basket" and benefit as much as possible from the complementarities offered by the range of activities available to farmers. This, as we shall see, is a considerable asset in terms of resilience and adaptive capacity. When other activities are accessible to the farmer or to one of his or her family members, these activities can also contribute to strengthening the household economy, provided that they are compatible with the maintenance of agricultural activity. In this case, we will use the concept of activity system to account for this phenomenon, understanding that this concept refers to the same scale of analysis as that of the production system, that of the household.[2]

These two scales of analysis alone, that of the cropping/livestock system and that of the farming/activity system, are, however, insufficient. Production units are indeed stakeholders in a territory; their workers have shaped the rural landscape. They themselves are inserted in complex social relations, particularly with regard to access to resources, means of production, information, and markets.

I.2.3 Studying practices and how they are embedded in social relationships

Contrary to what appears from certain official documents dealing with the consequences of climate change on agriculture, adaptation is not only a matter of changing practices or adopting new techniques. In fact, we hypothesize that the conditions of access to productive resources facing farmers, such as farmland, irrigation water, means of production, and labor force, but also the conditions of access to the market and knowledge, contribute significantly to informing farmers' choices and thus explaining certain aspects of the productive combination implemented. However, these conditions of access depend on the social relations that govern their availability and distribution, sometimes at the village level, but more likely at the small region level. Therefore, these processes are understood on a larger scale than the farm and the individual farmer's choices: the agrarian system scale.[3]

It is at the agrarian system level that we can identify the different categories of farmers who live and work in the same environmental conditions. We then can understand why they are not armed in the same way to face climatic hazards. The differentiation of production systems within the same region, and therefore the inequalities in development that it reveals, explains why they have neither the same resilience nor the same capacity to react and adapt to the possible deterioration of environmental conditions. In this book, the greatest attention will therefore be paid to the processes of differentiation of production systems and the typological approach. Each category of farmer (or domestic group) in a comparable situation in terms of access to resources, means of production, and the market often implements a similar combination of cropping and livestock systems, resulting from the past trajectory it followed, and is therefore in a comparable situation in terms of vulnerability and adaptation. We will try to highlight these trajectories in each of the regions studied, in particular those that explain the processes of impoverishment and increased vulnerability, but also those that, on the contrary, have created dynamics that limit the extent of the risks incurred.

Finally, farmers in a given region are also affected by national-level policy choices and conditions of integration into national and international trade through the prices at which they can expect to sell their production and purchase their inputs and equipment but also sometimes their food. This is why we will also focus on the analysis of past and current agricultural policies.

I.2.4 Farmers' past experiences

Climate change is nothing new, as the elderly farmers of the Sahel and the Sahelo-Sudanese regions know. They all remember the early 1970s, when a prolonged drought plunged many of them into distress. Far from being a limited episode, this drastic drop in rainfall (-30%), in a climatic context that was already tense in "normal" times, caused the entire geoclimatic band to experience a lasting climate deterioration. This is not the first time that farmers in these regions have had to fight against climate change, and they have extensive experience in how to deal with it.

This is why we thought it would be particularly interesting to give pride of place to these testimonies and reproduce all their richness. How were these difficulties lived? Who were those who managed to limit the impact and adapt to these new conditions? Which groups were the most vulnerable? And what were the effects of the agricultural policies implemented in these countries? How can we distinguish between the impact of the deteriorating climate and the impact of public policies?

The hindsight we can have on some of these past changes made by farmers themselves and their long-term consequences will allow us to glimpse the possible effects of changes implemented by farmers today, the conditions of their implementation, and foreseeable future consequences.

With even more hindsight, the interest shown in farmers' past experiences will also allow us to approach the distinction often made in current adaptation literature between, on one hand, "incremental" adaptations (also called "autonomous adaptations") and their effects, which are only perceptible over a long period of time,[4] and, on the other hand, "planned" larger-scale adaptations, generally by actors from outside the rural world (public authorities, projects, agricultural entrepreneurs, etc.), that, according to their promoters, are likely to bring about real systemic change toward greater adaptation. This will also allow us to better assess the impact that certain more "voluntary" adaptation projects or programs could make if they were implemented (Crane *et al.*, 2011).

I.3   Survey Method and Data Set

Since it was not possible at the chosen scale of analysis (the small agricultural region) to work from a corpus of reliable statistical data, the agrarian systems studies on which our analysis was based were centered on the collection of information through in-depth surveys with rural people: active farmers, elderly people, technical personnel, traders, and so on. These interviews and farm visits were sufficiently in-depth and repeated to be able to collect reliable first-hand information and gain a detailed understanding of the functioning of the various production systems

identified, from a point of view both technical and economic and social. The aim was also to take as much care as possible to understand the existing social relationships and their evolution, particularly with regard to access to resources (land, water, capital, and markets), as well as identify the impacts on local agrarian dynamics, national policy measures, development projects, population growth, market evolution, and, of course, possible climate change.

In each region studied, the fieldwork was supervised by one of us and carried out by young researchers enrolled at AgroParisTech in the *Comparative Agriculture Teaching & Research Unit*, either in their final year of engineering school and specialized in *Agricultural Development* or in their second year of the *Dynamics of Emerging and Developing Countries* (DynPED) master. The period spent in each "field," as close as possible to the farmers, was a full 6 months, all students having been supported on site by at least one supervision mission carried out by us halfway through the fieldwork. All interviews were conducted individually or in small teams, without the use of third-party interviewers. Great care was taken in the choice and sequence of questions asked, the simultaneous (and sometimes participatory) observation of people, tools, equipment, and machines, as well as gestures and their sequence.

In terms of surveys in rural areas, our working method is therefore similar to an ethnographic type of survey in that it involves observation, listening, immersion in the environment, daily note taking, and the absence of a closed-ended questionnaire. It is also similar in terms of its duration, the necessary contextualization of the survey, and its non-delegation to a third party; but on the other hand, it differs in terms of the requirement for purposive sampling and its concern for quantification (Cochet, 2015). By quantification, we mean obtaining *in fine* quantified and comparable results (production, costs, quantity of work, value-added, income, etc.) and not the use of quantitative methods of data processing.

In each region studied, we proceeded with a purposive sampling of the production units to be studied, a sampling based on the prior identification of the main "types" of production units present—thanks to the analysis of their landscape and the reconstitution of their past trajectories. The objective of this sampling was to reflect the diversity of agricultural production systems, allow the modeling of their technical functioning and their technical-economic results, and facilitate their comparison. The sample size was limited, with each production unit selected giving rise to a detailed case study. Thus, about 50 production units[5] were examined in each field, a sample to which one should add the historical surveys conducted with the elderly and, if necessary, other rural actors.

As for the study of older periods, particularly those in which farmers were already confronted with episodes of climatic deterioration and whose memory remains vivid, we proceeded by direct interviews with the elderly.[6] However, "talking about the weather" often results in too many stereotyped answers hastily given to the interviewer. "Climate change" is on everyone's lips, even when it is not clear what change is being talked about or whether it is really happening. The simple mention of "climate change" often opens the way to all sorts of considerations in which it is very difficult to sort out what is based on real observations, what is based on "feelings," and what is based on speeches heard here and there. This is why we have chosen to "study climate change without talking about it,"[7] or rather to analyze change through practices rather than through climate. Questioning the elderly about their past practices, taking care to situate what we are talking about in time and space, allows us to address— but only at a later stage—the issue of climate change. If there is a real change and if the farmers have felt the effects of these climate changes in one way or another, then they talk about it on their own without having to be asked. Subsequently, the possible associations of ideas between changes in practices and climate become much more significant and educational. They are much less indicative of fads or the desire to provide the interviewers with the answers they want to hear.

I.4 Selection of Fields Studied and Work Outline

A small number of contrasting situations have been chosen because they illustrate or demonstrate the questions raised by the consequences of climate change on agricultural systems and the resulting adaptation processes. They are far from being exhaustive in terms of diversity; many other case studies could have—should have—complemented our analysis. Moreover, the choice of the 12 regions studied was also made considering concrete possibilities for carrying out fieldwork: institutional contacts and the possibility of hosting young researchers in the field, and the security conditions required.[8] The majority of the regions studied (10 sites) are located in sub-Saharan Africa (see Fig. I.1), with four in

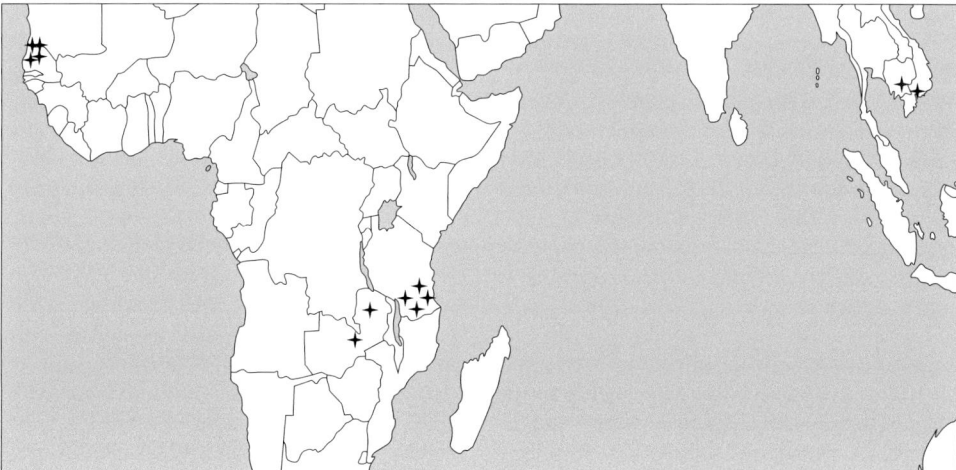

Fig. I.1. Location of the 12 regions studied.

the Sahelo-Sudanian zone (Senegal) and six in Southern Africa (Zambia and Tanzania). Two other regions are located in Southeast Asia (Cambodia and Vietnam); this choice allows us to increase the diversity of the cases studied while reinforcing the comparative character of our approach.

The first part of this work will address the case of sub-Saharan rainfed agriculture. Chapter 1 will deal with the situation of the peasantry in the Sahel-Sudan belt, based on two case studies in central and northern Senegal: the region of Bambey, located in the northern part of the peanut basin, and the region of Louga, chosen for its even more northern position (Cayor). These fields offer the opportunity to look at past episodes of climatic degradation (1970s and 1980s)—particularly severe in this region of the world—and the lessons to be learned in terms of vulnerability and adaptation. We will see that certain regions or social groups, although deprived of any possibility of developing irrigation, have nevertheless overcome these difficulties, at least in part, by adapting their practices and activity systems to these new constraints. In Chapter 2, we will discuss the agrarian systems in the highlands of Eastern and Southern Africa. We will draw on the study of two small regions in northern Zambia, Katongo Kapala in the District of Mpika and Miloso in the District of Mkushi (Fig. I.1), as well as a study of the Iringa region in the southern highlands of Tanzania. These three regions are characterized by rainfed agriculture in which corn (as a sole or associated crop) occupies a prominent place as both a food crop and a cash crop. This corn hegemony, as we shall see, contributes to making these agrarian systems highly vulnerable to possible changes in the rainfall pattern. Small-scale irrigation allows families who have access to it to diversify their production system and be partially freed from the rains' vagaries. The detailed study of these three regions will therefore make it possible to address both the question of vulnerability in rainfed agriculture in this region of the world and that of small-scale irrigation as a possible way of developing this agriculture, making it less subject to hazards while allowing one to anticipate the potential climatic deterioration predicted in the long term.

The second part of the book will be devoted to flood-prone regions where rice cultivation plays a major role in production systems. Chapter 3 will focus on two regions in southern Tanzania characterized by uncontrolled flooding with a complete lack of planning: first, the lower Rufiji River Valley, a region characterized by a complex ecosystem and development patterns that are both highly dependent on the river regime (major riverbeds, flood plains areas, etc.) and highly dependent on the rainfall regime on the hillsides; second, the Ifakara region on the Kilombero, a tributary of the Rufiji River. Here, farmers are confronted by two types of hazards, both linked to the climate. First, rainfall volume and distribution (date of the first significant rainfall, distribution of subsequent rains) and, second, flooding volume and timing (positioning, duration), on which the areas sown and harvested, the effectiveness of fertilization (deposited by the flood), and the yields obtained all depend. These regions are also part of the development "corridor" advocated by SAGCOT (Southern Agricultural Growth Corridor of Tanzania), where different models of agriculture (family farming and large-scale investment projects, including planning, particularly for rice cultivation) coexist. In the following chapter (Chapter 4) two contrasting situations will be compared. First, the Kampong Thom region on the banks of the Tonle Sap in Cambodia, where rice farmers take advantage of the Tonle Sap's rising water levels and the time lag between water accumulated from local rainfall and flooding to develop a complex rice-growing calendar, with two rainy season crop cycles, a flood recession cycle, and two (irrigated) dry-season cycles. Second, the Mekong Delta (in Vietnam), a region that could suffer from a likely significant increase in rainfall in its watershed, more violent floods, and a higher frequency of flooding. We will show how, at the village level, different socioeconomic groups have nevertheless managed to increase their standard of living, notably thanks to broad access to hydraulic planning, equipment, and facilities, although the delta remains subject to the major risk of rising sea level and the possible increased magnitude of floods. Chapter 5 will be devoted to the Senegal River Delta where two other small regions were studied. Low rainfall totals and interannual variations in the river's flooding explain farmers' and herders' long-standing adaptation to particularly constraining and uncertain climatic and hydrographic conditions in this region. Hard hit by the drought of the 1970s and 1980s, the region has also undergone profound changes under the influence of large-scale hydro-agricultural planning. We will thus have the opportunity to study the role that this planning has played in the adaptation of Upper and Lower Delta agriculture to the climatic upheaval at work, as well as the development models that they have brought forth.

In the third part (Chapter 6), we will look at the question of adaptation in mountain regions, where farmers must deal with altitude, slope, and the erosive phenomena it facilitates, stories, and sometimes communication difficulties. The Uluguru Mountains (in Tanzania)—the only example of a mountain region dealt with in this book—offer us a rich example of lessons learned, with steep slopes entirely cultivated by a very dense agricultural population (250–350 inhabitants/km²), systematic planning of the hillsides, generalized irrigation, and diversity of production.

Finally, in the book's fourth part, we will attempt to summarize the lessons to be learned from these few contrasting case studies. In Chapter 7, we will present the elements that seem to be the most decisive in farmers' ability to cope with hazards and to strengthen their capacity to adapt. We will emphasize the inequalities in development that characterize these agricultures, as well as their internal socioeconomic differentiation and its consequences in terms of adaptation. The last chapter (Chapter 8) will address the issue of policies to be promoted for Southern agriculture's adaptation to climate change. An overview of the policies generally promoted—or in the process of being promoted—will be presented via the analysis of a few reference texts from international or national institutions. Then, we will insist on the key attention points that, in our opinion, should underlie the development of adapted policies in this area.

Notes

[1] In this book, we employ the concept of "production system" as commonly used in agronomic research in the French-speaking world, rather than that of "farming system," which is widespread in English-language agricultural research (Fresco, 1984). "Farming system" can be used indiscriminately to refer to both the operation of a single farming production unit (with the aim of personalized advice) and that of a set of relatively homogenous farming production units. However, we have opted for the term "production system." It refers to a group of farms that fall within the same range of resources (in terms of surface area, equipment, and laborers), have comparable socioeconomic conditions, and practice a similar combination of productions; in short, a group of farms that can be represented by the same model (Cochet and Devienne, 2006; Cochet, 2015).

[2] On the methodological aspects of this approach, we refer the reader to the work carried out in the Comparative Agriculture Research & Training Unit of AgroParisTech. See in particular Cochet and Devienne (2006) and Cochet (2015).

[3] On the concept of agrarian system and its recent developments, see Cochet (2012).

[4] In the 2014 IPCC report, the editors define them as follows: "Autonomous adaptations are incremental changes in the existing system including through the ongoing implementation of extant knowledge and technology in response to the changes in climate experienced. They include coping responses and are reactive in nature" (IPCC, 2014b).

[5] The problems posed by identifying the contours of the "unit of production" in situations where the units of production, consumption, residence, and accumulation do not coincide, which are numerous in sub-Saharan Africa (Gastellu, 1979) have been addressed on a case-by-case basis in the regional studies.

[6] Consultation of the bibliography, when available, made it possible to carry out certain cross-checks. For example, the long-term evolution of rainfall in the Sahelian and Sahelo-Sudanese regions is well documented, since the drought of the 1970s helped to draw the attention of public authorities and financial backers to these issues and establish specialized institutions (notably the AGRHYMET Regional Center, a specialized institution of the Permanent Inter-State Committee for Drought Control in the Sahel CILSS). However, long-term climate series that are sufficiently reliable and allow the identification of possible past periods of climate change are not readily available in all locations. Although data already processed on a regional scale are often easily accessible on the Internet, particularly on the World Bank website, it is essential to return to climate data collected as close as possible to the regions being studied, as the deviations observed in relation to regional patterns are often significant. This necessarily leads to the following

difficulties: difficulties in accessing existing data, data that are sometimes subject to payment, incomplete series, doubts about the reliability of the measurements made, and so on.
[7] We refer to the work of Smadja *et al.* (2015), who faced the same concerns we did about the possible direction of surveyed farmers' responses at the mere mention of climate change.
[8] Lack of security makes many countries in the Sahelian and Sahelo-Sudanian zones inaccessible to field researchers.

References

Adger, W.N., Huq, S., Brown, K., Conway, D. and Hulme, M. (2003) Adaptation to climate change in the developing world. *Progress in Development Studies* 3(3), 179–195.

Altieri, M.A. (2016) *Developing and Promoting Agroecological Innovations within Country Program Strategies to Address Agroecosystem Resilience in Production Landscapes: A Guide.* Global Environment Facility/UNDP, New York, 25 pp.

Barnes, J., Dove, M., Lahsen, M., Mathews, A., McElwee, P. *et al.* (2013) Contribution of anthropology to the study of climate change. *Nature Climate Change* 3(6), 541–544.

Becerra, S. (2012) Vulnérabilité, risques et environnement: l'itinéraire chaotique d'un paradigme sociologique contemporain. *VertigO* 12(1).

Becerra, S., and Peltier, A. (eds) (2009) *Risques et environnement: recherches interdisciplinaires sur la vulnérabilité des sociétés.* L'Harmattan, Paris, France, 575 pp.

Bidou, J.-E. and Droy I. (2009) Décrire la construction temporelle des vulnérabilités: Observatoires ruraux et analyse historique des moyens d'existence dans le Sud malgache. In: Becerra, S. and Peltier A. (eds) *Risques et environnement: recherches interdisciplinaires sur la vulnérabilité des sociétés.* L'Harmattan, Paris, France, pp. 155–170.

Boit, A., Sakschewski, B., Boysen, L., Cano-Crespo, A., Clement, J. *et al.* (2016) Large-scale impact of climate change vs. land-use change on future biome shifts in Latin America. *Global Change Biology* 22(11), 3689–3701.

Bonneuil, C. and Fressoz, J.-B., (2016) *The Shock of the Anthropocene: The Earth, History and Us.* Verso, London, 320 pp.

Brémond, P., Grelot, F. and Bauduceau, N. (2009) De la vulnérabilité de la parcelle à celle de l'exploitation agricole: un changement d'échelle nécessaire pour l'évaluation économique des projets de gestion des inondations. In: Becerra, S. and Peltier, A. (eds) *Risques et environnement: recherches interdisciplinaires sur la vulnérabilité des sociétés.* L'Harmattan, Paris, France, pp. 231–244.

Brown, K. (2016) *Resilience, Development and Global Change.* Routledge, London, 212 pp.

Buchheit, P., d'Aquino, P. and Ducourtieux, O. (2016) Cadres théoriques mobilisant les concepts de résilience et de vulnérabilité. *VertigO* 16(1).

Cash, D.W., Adger, W.N., Berkes, F., Garden, P., Lebel, L. *et al.* (2006) Scale and cross-scale dynamics: governance and information in a multilevel world. *Ecology and Society* 11(2).

Chew, S. and Sarabia, D. (2016) Nature–culture relations: early globalization, climate changes, and system crisis. *Sustainability* 8(1), 78–106.

Cochet, H. (2012) The systeme agraire concept in francophone peasant studies. *Geoforum* 43, 128–136.

Cochet, H. (2015) *Comparative Agriculture.* Quae/Springer, Dordrecht, Netherlands, 154 pp.

Cochet, H. and Devienne, S., (2006) Fonctionnement et performances économiques des systèmes de production agricole: une démarche à l'échelle régionale. *Cahiers Agricultures* 15(6), 578–583.

Crane, T.A., Roncoli, C. and Hoogenboom, G. (2011) Adaptation to climate change and climate variability: the importance of understanding agriculture as performance. *NJAS—Wageningen Journal of Life Sciences* 57, 179–185.

Da Cunha, C. and Vanderlinden, J.-P. (2014) Adaptation aux changements globaux: quel apport de la trans-disciplinarité? *Revue française de Socio-Economie* 204(13), 277–282.

Dupuis, J. (2015) *S'adapter au changement climatique: Analyse critique des nouvelles politiques de gestion de l'environnement.* Alphil Editions, Neufchâtel, France, 392 pp.

Fresco, L. (1984) Comparing anglophone and francophone approaches to farming systems research and extension. In: *Farming Systems Support Project Networking Paper* 1, 36 pp.

Gastellu, J.-M. (1979) Mais où sont donc ces unités économiques que nos amis cherchent tant en Afrique? Note de travail, Série: enquêtes et outils statistiques, vol 1, « Le choix de l'unité », AMIRA, pp. 99–122.

Hiwasaki, L., Luna, E., Syamsidik and Marçal, J.A. (2014) Local and indigenous knowledge on climate-related hazards of coastal and small island communities in Southeast Asia. *Climatic Change* 128(1–2), 35–56.

Huang, W.-C. and Lee, Y.-Y. (2016) Strategic planning for land use under extreme climate changes: a case study in Taiwan. *Sustainability* 8(1), 53–69.

IPCC (2014a) *Climate Change 2014: Impacts, Adaptation, and Vulnerability. Part A: Global and Sectoral Aspects. Contribution of Working Group II to the Fifth Assessment Report of the Intergovernmental Panel on Climate Change.* Field, C.B., Barros, V.R., Dokken, D.J., Mach, K.J., Mastrandrea, M.D. *et al.* (eds). Cambridge University Press, Cambridge, UK, 1132 pp.

IPCC (2014b) *Climate Change 2014: Impacts, Adaptation, and Vulnerability. Part B: Regional Aspects. Contribution of Working Group II to the Fifth Assessment Report of the Intergovernmental Panel on Climate Change.* Barros, V.R., Field, C.B., Dokken, D.J., Mastrandrea, M.D., Mach, K.J. *et al.* (eds). Cambridge University Press, Cambridge, UK, 688 pp.

Janin, P. (2006) La vulnérabilité alimentaire des sahéliens: concepts, échelles et enseignements d'une recherche de terrain. *L'espace géographique* 35(4), 355–366.

La Branche, S. (éd.) (2011) *Le changement climatique, du méta-risque à la méta-gouvernance.* Lavoisier, Paris, France, 220 pp.

Lallau, B. and Rousseau, S. (2009) De la vulnérabilité à la résilience: une approche par les capabilités de la gestion des risques. In: Becerra, S. and Peltier, A. (eds) *Risques et environnement: recherches interdisciplinaires sur la vulnérabilité des sociétés.* L'Harmattan, Paris, France, pp. 171–183.

Lallau, B. and Thibaut, E. (2009) La résilience en débat: quel devenir pour les agriculteurs en difficulté? *Revue d'Etudes en Agriculture et Environnement* 90(1), 79–102.

Leichenko, R. and Silva, J.A. (2014) Climate change and poverty: vulnerability, impacts, and alleviation strategies. *WIREs Clim Change* 5, 539–556.

Locatelli, B. (2016) Ecosystem services and climate change. In: Potschin, M., Haines-Young, R., Fish, R. and Turner, R.K. (eds) *Routledge Handbook of Ecosystem Services.* Routledge, London, pp. 481–490.

Mathevet, R. and Bousquet, F. (2014) *Résilience & Environnement: penser les changements socio-écologiques.* Buchet-Chastel, Paris, France, 176 pp.

Mullenite, J. (2017) Can climate change adaptation be a desirable goal? *Human Geography* 10(2), 87–94.

Poteete, A.R., Janssen, M.A. and Ostrom, E. (2010) *Working Together: Collective Action, the Commons, and Multiple Methods in Practice.* Princeton University Press, Princeton, New Jersey, 370 pp.

Reghezza-Zitt, M. and Rufat, S. (eds) (2015) *Résiliences: Sociétés et territoires face à l'incertitude, aux risques et aux catastrophes.* ISTE, London, 227 pp.

Rhodes, E.R., Jalloh, A. and Diouf, A. (2014) Revue de la recherche et des politiques en matière d'adaptation au changement climatique dans le secteur de l'agriculture en Afrique de l'Ouest. Document de travail 090, Future Agricultures, IDRC/CRDI, 56 pp.

Smadja, J., Aubriot, O., Puschiasis, O., Duplan, T., Grimaldi, J. *et al.* (2015) Changement climatique et ressource en eau en Himalaya. Enquêtes auprès de villageois dans quatre unités géographiques du bassin de l Koshi, Népal. *Journal de géographie alpine* 103(2): 23.

Springmann, M., Mason-D'Croz, D., Robinson, S., Garnett, T., Godfray, H.C.J. *et al.* (2016) Global and regional health effects of future food production under climate change: a modelling study. *The Lancet* 387(10031), 1937–1946.

Tasser, E., Leitinger, G. and Tappeiner, U. (2017) Climate change versus land-use change: what affects the mountain landscapes more? *Land Use Policy* 60, 60–72.

I

Sub-Saharan Rainfed Agriculture

_____

1 Agrarian Dynamics and Climate Change in the Senegalese Sahelian Peanut Basin

Nadège Garambois[1]*, Ulysse Le Goff[2], and Lucie Thibaudeau[2]
[1]*Comparative Agriculture Training & Research Unit/UMR Prodig, AgroParisTech, Palaiseau, France;* [2]*Agroeconomist, AgroParisTech, Palaiseau, France*

1.1 Introduction

Although the Sahel had already experienced episodes of severe drought in its history (1880, 1912–1914, 1940–1944; Jouve, 1991), the drought that began at the end of the 1960s was exceptional in terms of its persistence, intensity, and spread throughout Africa south of the Sahara. From 1968 to 1984, the 100 mm isohyet went down by 250–300 km to the south, and the 500 mm isohyet by 150 km (Le Borgne, 1988). Northern and central Senegal was particularly affected, with rainfall deficits of more than 50% in 1968–1973 and 1977–1984. Based on the study of the center and north of the Senegalese peanut basin, the purpose of this chapter is to analyze how farmers have been striving for a long time to limit their vulnerability under difficult and uncertain rainfall conditions, examine the impact of this long period of drought on farming households, and study how they have adapted.

Two study areas were selected in the historical regions of Cayor and Baol (Fig. 1.1):

- The Louga region (Le Goff, 2016) in the north of the peanut basin, which was already under a Sahelian climate (in the sense of the 500 mm annual rainfall isohyet limit) before the major drought of the 1970s–1980s and received below 300 mm of rainfall per year during that period

- The Bambey region (Thibaudeau, 2015) in the center of the peanut basin, which had a Sudanese climate in the first half of the 20th century but has had a Sahelian climate since the 1970s

In the absence of surface water, the progressive settlement of these two regions relied on the digging of artisanal wells several tens of meters deep for human and animal drinking water. Today, these wells are backed up by much deeper boreholes drawing on the salty Maastrichtian water table, whose water is reserved for non-food and non-agricultural domestic uses. The sudden and permanent deterioration of climatic conditions in the 1970s was coupled, particularly from the 1980s onward, with a profound overhaul of Senegal's agricultural policy, which affected the entire peanut basin and forced farmers to adapt in parallel to a profound economic upheaval.

---

*Corresponding author: nadege.garambois@agroparistech.fr

©2024 CAB International. *Agrarian Systems and Climate Change: Journeys of adaptation in the Global South* (eds H. Cochet, O. Ducourtieux, and N. Garambois)
DOI: 10.1079/9781800628137.0001

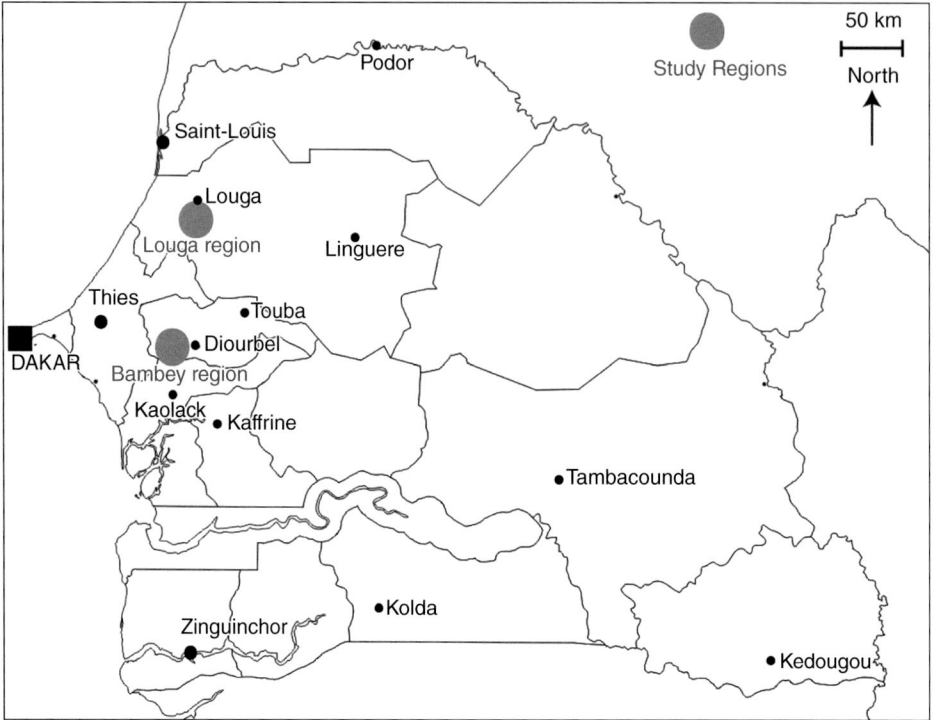

Fig. 1.1. Location of the Bambey and Louga regions in Senegal.

1.2 Rainfed Sahelian Agriculture Has Long Been Adapted to Climatic Hazards

1.2.1 Agriculture in the Sahelian Peanut Basin in the 1950s

Already under a Sahelian climate in the 1950s, the northern part of the peanut basin was home to rainfed agriculture based on ruminant livestock and millet and peanut crops grown in a regulated cropping pattern on parkland. Three years of crops were followed by 3–10 years of fallow land. In the first half of the 20th century, average annual rainfall in the Louga region was less than 500 mm, and the rainy season lasted less than 90 days on average. These climatic conditions only allowed for one rainfed crop cycle. Farmers used early varieties compared with the more southern and rainy regions of Senegal: *souna* millet, cowpea, and watermelon (70 days); *sanio* millet, peanut, and sorghum (90 days) (Chevalier, 1947; Portères, 1950). Given

the rainy season's very short duration and the strictly manual nature of this type of agriculture at the time, most of the cultures (millet, sorghum, and cowpea) were sown in dry soil, without waiting for the first rains. The aim was to limit the labor peak at the rains' arrival and allow all these crops to benefit from the entire rainfall duration. If the first rain was particularly late or the second rain too late (risking seed rot), the low seed weight per hectare for millet and sorghum made it possible to resow at lower cost, if necessary, preferably with the earliest varieties. Only peanut, whose seed stock is much smaller (large seeds and more expensive), was sown after the first rain. If the soil was sufficiently loose, working the soil with an *iler* at a rather shallow level (3–5 cm deep), or even simply direct seeding in a seed hole using a tool called a *larmet*, helped limit evaporation and accelerate the soil preparation and sowing stages.

Some agricultural production was sold (peanut for the colonial oil industry, livestock products), but surpluses were often small, and agriculture in the Louga region was centered on on-farm consumption. On the other hand, the robustness that this food autonomy conferred on families, in the absence of significant sources of extra-agricultural income, made most of them all the more dependent on their agricultural production. However, rainfall conditions as well as crop yields were already described as highly uncertain from one year to the next (Chevalier, 1947; Portères, 1950). In this context, staggering the harvest calendar contributed in particular to reducing families' food vulnerability by limiting the length of the bridging time. *Souna* millet, cowpeas, and watermelons (70-day cycles) took over the diet as soon as harvested in September. *Sanio* millet, sorghum, and peanut were harvested after the end of the rainy season (late October–November); millet and sorghum were stored and consumed during the dry season. The cassava crop played a major role in this regard. Cultivated on lowland fields, in more clayey soils and with higher soil water reserves, cassava was the only crop similar to an off-season crop. Its deep rooting (up to 1-m deep) allowed this crop to explore deeper horizons, while the greater proximity to the water table allowed by the topography facilitated capillary rise. Harvested as early as April and spread out thanks to being well preserved in the soil, cassava was a valuable fallback in the event of poor millet and sorghum harvests. Orchard gardens installed near houses supplemented food resources, baobabs, and mango trees being particularly useful (if not vital) in years when the bridging time was more difficult.

Soil fertility maintenance was based on the progressive construction and upkeep of the parkland associated with the crops in cultivated areas. It was composed of *Faidherbia albida* (*kad*), *Acacia raddiana*, *Balanites aegyptia* (*sump*), and *Neocarya macrophylla* (Gingerbread plum) and thus included leguminous plants and species useful for gathering. *F. albida* played a major role in this respect by providing mineral elements through vertical transfers from:

- deeper soil horizons, thanks to the elements and products of substrate alteration drawn by its root system; and
- the atmosphere, thanks to the nitrogen fixation by the symbiotic bacteria installed in nodosities on its root system.

The vegetative rhythm of *F. albida* made it non-concurrent with rainfed annual crops; the full development of its foliage in the dry season was made possible by drawing water from deep within the soil, at a time when the soil water reserve from the surface horizon was too weak for rainfed crops. Its shading limits soil evapotranspiration. At the end of the dry season, fallen leaves and pods considerably enrich the soil in minerals and carbon, while not shading annual crops in the rainy season. Moreover, by consuming these pods, cattle facilitate the germination of their seeds, which come out in livestock waste.

The restoration and transfer of fertility were also based on the close linkage between crops and livestock. Indeed, the herbaceous fallow land following cultivation years was grazed, within the framework of a regulated plot allotment during the rainy season. All cultivated land was then dedicated to common grazing in the dry season (with the exception of sorghum fields, which were protected by hedges), with animals' night resting on the fields that would be cultivated the following year in order to concentrate fertility transfers. In addition to sedentary farmers' herds, the region was a transhumance site for herders from the Ferlo and their herds in the dry season. This large-scale mobility of their animals, and those entrusted to them by farmers, was a key factor in the adaptation of livestock to a variable forage resource in time and space. The bartering of cereals and milk accompanied the cohabitation of mixed crop and livestock farmers and transhumant livestock herders for several months and reinforced the nutritional robustness of this agrarian system, which was subject to difficult and unpredictable rainfall conditions.

Further south, the agrarian system in the Bambey region, in the center of the peanut basin, was based on the same production and a similar organization but benefited from a Sudanese climate and much higher rainfall. Farmers also used early varieties of *souna* millet for food security (but here the cycle reaches 90 days) and could grow semi-late varieties (120 days) of *sanio* millet, peanut, and sorghum. In the 1950s, these varieties were sowed at the first rains and not when the soil was dry, as in Louga. At the time, the yields recorded for crops and fodder available on fallow land and rangeland were much higher than those in the Louga region.

1.2.2 The effects of the state-supported agricultural revolution in the 1960s and 1970s

Senegal's independence in 1960 and Léopold Sédar Senghor's accession to power were accompanied by agricultural policies through the Agricultural Program (1960–1980). This program aimed to increase agricultural production, particularly peanuts, the main source of foreign currency for the country, by using fertilizers and selected seeds and accelerating the allocation of equipment to Senegalese farmers.

Throughout Senegal, 226,000 seed drills, 294,000 cultivators and 71,000 "lifters" for digging up peanuts were distributed over 20 years. The promotion of this equipment was accompanied by a subsidized fertilizer credit policy and a tenfold increase in fertilizer consumption over this period (Freud *et al.*, 1997).

Created at the dawn of independence, the *Office de Commercialisation Agricole* (Office of Agricultural Commercialization), or OCA, has been in charge of purchasing crops and delivering and distributing seeds since 1960, while the *Centres Régionaux d'Assistance au Développement* (Regional Centers for Development Assistance), or CRAD, created in 1964, encouraged and supported the formation of cooperatives where farmers acquired equipment and inputs and sold their production. In 1966, the creation of the *Office National de Coopération et d'Assistance pour le Développement* (National Office of Cooperation and Assistance for Development), or ONCAD, which combined the functions of the OCA and the CRAD, established the state's monopoly on the peanut and fertilizer trade (Dieng and Gueye, 2005).

The policy of guaranteed peanut prices was a continuation of the end of the colonial period. Since 1933, Senegalese peanuts had benefited from a preferential tariff, thanks to the establishment of a customs duty on peanuts exported to France. The revenue was used to subsidize the peanut purchase price for producers. The strengthening of the Agricultural Program in 1954 contributed to a significant increase in Senegalese production. In 1960, the subsidy still represented 30% of the FOB (Free on Board) price (Freud *et al.*, 1997).

In the regions studied in the north and center of the peanut basin, the Agricultural Program was reflected in two campaigns to distribute equipment (seed drills, cultivators, carts) to villages where cooperatives had been organized. In theory, all farmers had access to this equipment in exchange for a payment of 35 kg of shelled peanuts per year for 5 years. However, the equipment made available was not sufficient to satisfy all the demand, nor to equip all the families. Overall, the allocations favored farmers and villages involved in cooperatives and those who produced enough peanuts to make a 5-year commitment and pay their debts to the cooperative.

Some families thus had all the necessary equipment for draft power (seed drill, cultivator, cart) by the mid-1960s. Other families were able to acquire one or two of these three tools and exchange equipment in order to have access to the missing ones, while the poorest families were totally dependent on borrowing the equipment.

However, in response to demand, blacksmiths took over by making not only hand tools but also animal draft power equipment based on the tool models distributed under the Agricultural Program. Although these handmade tools were lighter and less resistant, they were also more accessible financially and allowed families who did not have sufficient cash or the necessary connections with the cooperative leaders to be adequately equipped in the early 1970s. The dissemination of this animal draft power equipment went hand in hand with the dissemination of selected peanut seeds with an erect growth habit (rather than the creeping variety) in order to mechanize weeding more widely.

In order of priority and according to their means, farmers chose to equip themselves with cultivators, which allowed them to reduce the major labor peak presented by manual hoeing, and then, if their investment capacity allowed, with a seed drill or even a cart. For peanut cultivation, labor peaks corresponded to weeding and harvesting periods, and much less to soil preparation or sowing periods. Thus, for peanuts, 45 man-days (MD) of manual labor per hectare were sufficient for preparing and maintaining the crop until harvest. In contrast, the equivalent of more than 80 days of work for women and children were required for the delicate task of pod removal (Chevalier, 1929; surveys). In the 1970s, blacksmith-manufactured draft animal "lifters" to quickly dig up peanuts helped increase the area that a family could harvest. The fertilizers sold and the equipment distributed primarily concerned peanuts, but productivity gains have also been recorded for cereal crops—millet, grown in rotation with peanuts, benefits from the carry-over effect (of peanut crop). Thanks to the labor productivity gain, the families equipped with at least one cultivator (the animal draft "hoe"), or which could borrow one, were able to increase their cropping area. The diffusion of animal draft power (Photo 1.1) thus led to a large extension of sown land to the detriment of fallow land—3 years of cultivation followed by a fallow period of no more than 3 years in Louga and continuous cultivation in Bambey—and of *ager* to the detriment of *saltus*. The latter no longer allowed for the feeding of herds on-site during the rainy season. As a result, herds were driven to transhumance in the Ferlo during the rainy season but upon their return they had abundant crop residues left in the field that were consumed as common pasture during the dry season. Only peanut stalks were harvested more often to feed horses and donkeys, as these animals were now used not only as pack animals but also for draft.

In the regions studied, social differentiation between families was expressed primarily in manual farming, by the size of the family and the size of the herd, which respectively controlled the area that could be worked and fertilized. From that time on, it was based on differences in equipment levels between families with a full range of equipment and draft tools (and cash flow that allowed for the broad use of fertilizer and selected peanut seeds) and those that still cultivated with strictly manual tools.

Increased labor productivity (yield and area cultivated per farmer) and favorable price ratios contributed to a veritable agricultural revolution in the Senegalese peanut basin in the 1960s. This period is described as a golden age by the farmers who experienced these transformations. However, this revolution was only possible at the cost of frequently forgoing the repayment of equipment loans; the increase in production and the high guaranteed peanut prices were not, for many producers, sufficient to allow them to accumulate the necessary capital. These changes and the increased specialization in peanut production also reflect how these technical developments and producers' income depended on public support and this export crop.

1.3   The Double Shock of the 1970s and 1980s, Acute Drought, and Upheaval of the Agricultural Policy Framework

1.3.1   Manifestations and effects of the climate crisis of the 1970s and 1980s

1.3.1.1   Climatic and hydric manifestations of drought

The north and center of the peanut basin experienced a profound change in rainfall from the end of the 1960s until the end of the 1980s and even the 1990s. Various parameters related to rainfall had a particularly central influence on agricultural practices in the Sahelian zone: the length of the rainy season, the date of the first rains, the number and distribution of rainy days during the season, and the frequency of the various dry and wet episodes during wintering. Louga, which already had a Sahelian climate in the first half of the 20th century, saw a drop in average rainfall from 470 mm between 1946 and 1967 to 270 mm per year from 1968 to 1986 (a decline of nearly 45%, Fig. 1.2), placing this region at the limits of a Sahelo-Saharan climate (below 250 mm of rain per year) and the maintenance of rainfed agriculture. According to data from the Directors of Statistical Analysis and Prediction in Dakar, the duration of the rainy season, averaging 80 days over the 1961–1967 period, fell to less than 50 days over the 1968–1998 period (Sambou *et al.*, 2015).

Although starting from a significantly higher average level (700 mm of rainfall per year over the 1946–1967 period), rainfall in the Bambey region also declined sharply (35%), falling to an average of 450 mm per year between 1968 and 1986 (with many years around 350 mm), and causing the region to fall into a Sahelian climate (Fig. 1.3). According to data from the *Agence Nationale de l'Aviation Civile et de la Météorologie* (National Agency of Civil Aviation and Meteorology), or ANACIM, this sharp decline in rainfall was accompanied by a 1-month shortening of the rainy season: 105 days on average from 1946 to 1967, compared with 75 days on average from 1968 to 1999.

The induced lowering of the water table reduced the capillary rise of moisture to the superficial horizons of the soil, a phenomenon

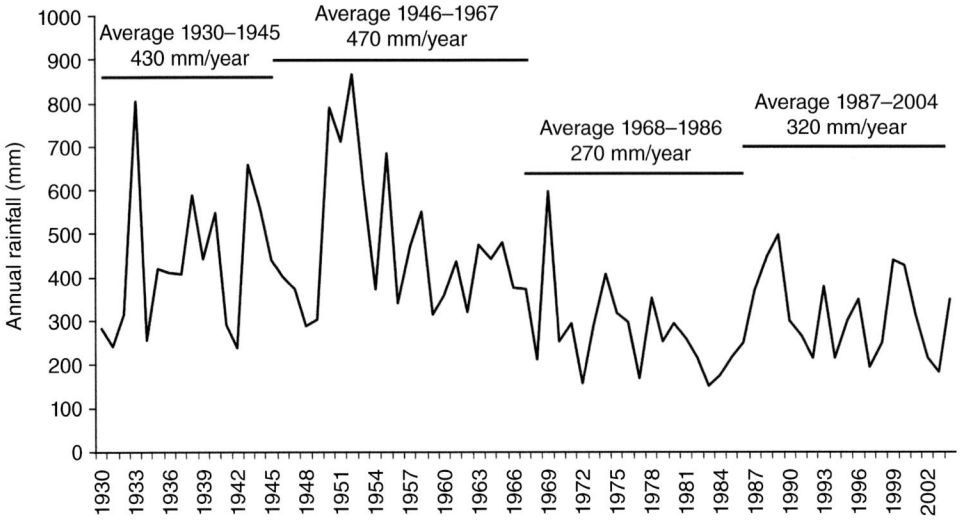

Fig. 1.2. Evolution of annual rainfall in Louga from 1930 to 2004 (Le Borgne, 1988; ANACIM data).

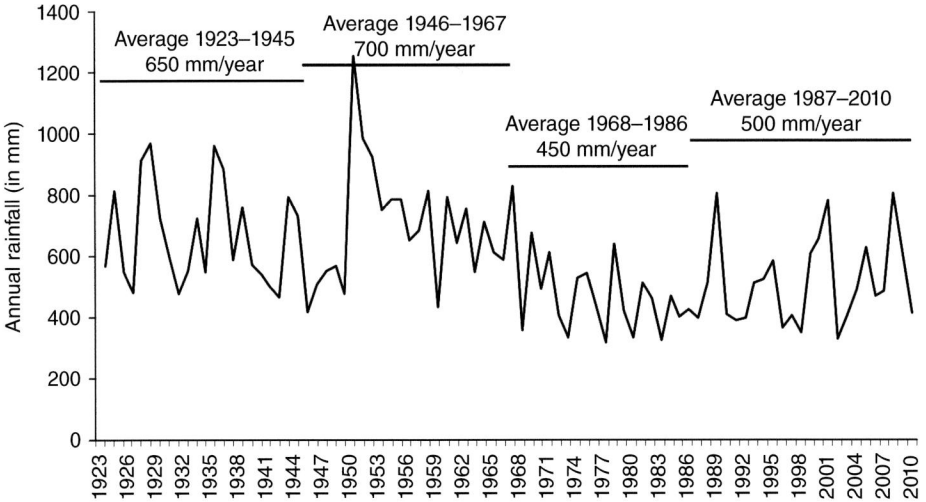

Fig. 1.3. Evolution of annual rainfall in Bambey from 1923 to 2010 (Le Borgne, 1988; ANACIM data).

that favors tree species whose root systems are capable of exploring slightly deeper horizons. The general decrease in soil moisture in the dry season, particularly in the superficial horizons, made them all the more sensitive to wind erosion.

1.3.1.1.1   VERY SCARCE HARVESTS AND BRUTAL DE-CAPITALIZATION.   The drop in rainfall totals in the Louga and Bambey regions and the interannual irregularity of rainfall conditions (particularly in terms of the number of days of rain) complicated

crop management and the organization of the cropping calendar (Roquet, 2008). If the first rainfall is not followed by sufficiently regular showers, sowing must be repeated. Dry spells in the middle of the rainy season occurred with greater frequency, and at a stage when plants were particularly sensitive to water stress (flowering and maturation in particular), leading to a sharp drop in yield.

The varieties grown in each of the two regions until the end of the 1960s, which were adapted to a longer rainy season, proved to be unsuited to these new climatic conditions: the growth and maturation cycle could no longer be completed, leading to a drop in yields in years with a high-water deficit. Chalby and Demarly (1991) report that an early peanut variety (90 days) that completed its cycle in Louga 14 out of 17 years from 1953 to 1969, was only adapted 3 out of 17 years from 1970 to 1986. At the same time, the rains' more erratic nature at the end of wintering led to concerns about the risk of germinating peanuts in the soil before harvest. The cultivation of cassava, which played an important role in the bridging time in the Louga region, became impossible. The situation was less catastrophic in Bambey, where rainfall and the length of the rainy season allow the use of early varieties of peanuts, sorghum, and millet (90 days maximum) that were previously cultivated in more northerly regions. Yields, however, were much lower and were weakened by possible drought periods during cultivation (Chalby and Demarly, 1991).

Access to animal draft power, a legacy of the 1960s–1970s Agricultural Program, and sowing millet in dry conditions, which allowed germination to begin on all fields as soon as the first rains fell (this practice came back to the Louga region, and was new to the Bambey region), were decisive factors in the tightening of the calendar window for all cultivation operations (sowing, weeding, harvesting). They allowed for a much larger area to be cultivated per farmer than would be possible with strictly manual tools, and then for a less small harvest provided that rainfall conditions allow the entire crop cycle to take place.

Despite the food aid put in place in the 1970s and 1980s (Minvielle and Lailler, 2005) and the evolution of equipment, the excessive repetition of sometimes extremely poor harvests did not allow families to survive without having to decapitalize, most often by selling some of their animals and, in the most extreme cases, their equipment. Surveys conducted in the peanut basin by Auserehl Kelly (1989) at the time indicate that more than 70% of the producers surveyed were unable to meet their basic needs in years of poor harvests (e.g. 1983–1984 campaign).

Faced with declining forage resources throughout the Sahelian zone, transhumant herders, their herds, and the cattle and small ruminants entrusted to them, spent less time in the northern part of the peanut basin during the dry season and moved more quickly to the central region. During the rainy season, they move to the "Terres Neuves" region in the southeast of the country, which had more rainfall and fewer people. The animals' undernourishment, which made them all the more vulnerable to disease, led to epizootic diseases that decimated the herds. Transhumant herders saw their livestock dwindle. In the regions of Bambey and Louga, very few farming families still had a few head of cattle at the end of the 1980s.

1.3.1.1.2 LABOR MOBILITY AND OFF-FARM INCOME IN THE FACE OF CRISIS IN THE NORTHERN PEANUT BASIN. Faced with this crisis, families in the northern part of the peanut basin decided in the 1970s to send some of their young people to the city during the dry season. In the Louga region, for example, the most affluent families sent these young people to train in the city so that, after 1 or 2 years, they could take up a more lucrative skilled activity. In return, these young people could contribute, through money transfers, to the lives of extended family members who remained in the village. During this period of very poor harvests, this off-farm income was crucial for the survival of rural families.

1.3.1.2 Progressive weakening of the Senegalese peanut market and a break in agricultural policy at the turn of the 1980s

At the same time as this climatic crisis, Senegalese agricultural policy underwent profound upheavals and scrutiny of the support mechanisms for peanut cultivation from 1983 onward, which further weakened producers in the peanut basin.

1.3.1.2.1 AN EEC COMPENSATORY SCHEME LESS ADVANTAGEOUS FOR SENEGALESE PEANUTS THAN THE PREFERENTIAL TARIFFS GUARANTEED BY FRANCE UNTIL 1965. Until 1960, the organization of the peanut market in the franc zone was based on import controls, facilitated by the low degree of substitution of fats on the French market, where peanut oil predominated. It ensured a guaranteed price for peanut producers and outlets for oil producers. This organization, renewed during the first postindependence campaign by an agreement between Senegal and France, was called into question by the latter's membership in the European Economic Community (EEC). The second association convention between the *États Africains et Malgaches Associés* (African and Malagasy associated states), or EAMA, and the EEC member countries, signed in Yaoundé in 1963, came into force for Senegal in 1966. It marked the beginning of the establishment of a free trade area between the EAMA and the EEC (Debien, 1966).

At the same time, the French oilseed market, until then protected by restrictions on fat imports from third countries, was gradually opening, like that of its community partners, to the circulation of competing products. The Dillon Round negotiations within the framework of the GATT (General Agreement on Tariffs and Trade) marked the beginning of oilseeds' entry into the European market, without customs duties. The intense processes of substitution between products, which characterize the international fats market, put Senegalese exports in full competition with the United States on the community market, forcing them to align with world prices.

A compensatory mechanism foreseen by the EEC since the mid-1960s and specific to EAMA oilseed exports should have ensured that the latter benefit from a stable "minimum world price" that was higher than the international market price, based on the quantities imported by the EEC. This system attempted to compensate, to some extent, for the decline in world prices. It was based on a fixed annual envelope taken from the EAGGF (European Agricultural Guidance and Guarantee Fund), which did not allow partner countries to increase their production unless they accept a minimum price closer to the world price (Debien, 1966). After 1975 and the signing of the first Lomé Convention, the system was integrated into the Stabex system, which sought to offset, through financial compensation, the alignment of the 46 ACP (African, Caribbean, and Pacific) countries with world prices in the European market for many tropical products. After an initial boost to exports from ACP countries, the drop in world prices forced a ceiling on the Stabex envelope to prevent the system from going bankrupt and reduced its impact on producers' incomes (Ministère des Relations extérieures, de la Cooperation et du Développement, 1983).

1.3.1.2.2 FALLING PEANUT PRICES, INABILITY TO REPAY AGRICULTURAL LOANS, ONCAD'S BANKRUPTCY, AND CESSATION OF INPUT SUBSIDIES. Designed based on French and European market organization as it existed in the early 1960s (guaranteed prices and outlets), the Agricultural Program was maintained until 1980, despite the profound changes in the Senegalese peanut trade that were unfavorable to both Senegalese producers and the balance of the state budget through ONCAD. Faced with the unfavorable evolution of price ratios for producers, many credits granted for seed and equipment purchases were not repaid. In addition to the debts accumulated by ONCAD and the second oil crisis, this non-repayment led to the bankruptcy of this state office, which had a monopoly on the peanut and fertilizer trade in Senegal (Freud *et al.*, 1997). Its dissolution and the termination of the Agricultural Program, replaced by the Economic and Financial Recovery Program (PREF) from 1981 to 1985, led to a drying up of credit for producers' fertilizer purchases and a sharp drop in consumption (to zero in 1983) (Dieng and Gueye, 2005) in the midst of a rainfall deficit.

1.4 Adaptation of Agriculture to Rainfall Deterioration in a Context of Government Disengagement and Demographic Growth

1.4.1 Reinforced structural adjustment from 1995 (Pasa)

In response to nearly two decades of undermining of the Senegalese peanut market, the *Nouvelle Politique Agricole* (new agricultural policy), or NPA, and the Cereal Plan implemented from 1985 to 1994 aimed for a recovery of agricultural prices and the protection of local cereals through a state monopoly on rice imports and commercialization (Dieng and Gueye, 2005). The fixed prices practiced until then were replaced by price ceilings and floors, while at the same time uncontrolled cereal markets were maintained. In fact, the large-scale targeted actions primarily concerned the development of the Senegal River Valley and national rice production, and much less agriculture in the peanut basin, which received only 7% of the agricultural investment program package from 1987 to 1995 (Oya, 2009). At the same time, the state's role in marketing cereals, supplying inputs (seeds and fertilizers), and providing access to credit was handed over to the private sector, which was accompanied by the elimination of input subsidies (Dieng and Gueye, 2005).

The state's disengagement in terms of access to inputs was accompanied by a relative evolution of prices that was very unfavorable to farmers (Fig. 1.4).

1.4.2   Slight increase in rainfall, but strong hazards remain

The rainfall deficit observed in Senegal from the late 1960s to the late 1990s was slightly reduced (in the range of +10% to +17% of rainfall on average over the area). However, the first rain could be increasingly comparable to "false starts" because they are followed by fairly long rainfall breaks, sometimes exceeding 2 weeks. Rainfall interruptions of more than 7 days thus persist until the first 10 days of July in the central regions (and until the last 10 days of July in the northern regions of Senegal), while presenting a higher occurrence in the north and northcentral

regions of the country. The start of the rainy season is then characterized by heavy rainfall at the beginning of the season. Isolated from the rest of the rainfall, it can cause severe losses on seedlings (Salack et al., 2012). While slightly wetter years may be on the horizon, temperatures are still slowly but steadily increasing, which contributes to increased evaporation.

In Bambey, for example, since 1987 annual rainfall has fluctuated between 360 and 800 mm per year, with an average of 505 mm per year over the 1987–2010 period. Although the average exceeds the 450 mm per year recorded during the 1968–1986 period when the drought was most severe, the previous rainfall level has not been recovered (Fig. 1.3). Rainfall remains reduced during the 1990s and early 2000s with the rainy season still reduced to two and a half months. Despite the recovery that seems to be under way (average rainy season length of 85 days over the 1999–2010 period), strong interannual variability remains.

In the northern part of the peanut basin, after 2 years of severe drought in 1997–1998 (195 and 250 mm of rainfall per year in Louga, respectively), rainfall seems to be trending upward, although it has not returned to pre-crisis levels, nor has it escaped extreme years: 185 mm of cumulative rainfall and a 10-day rainy season

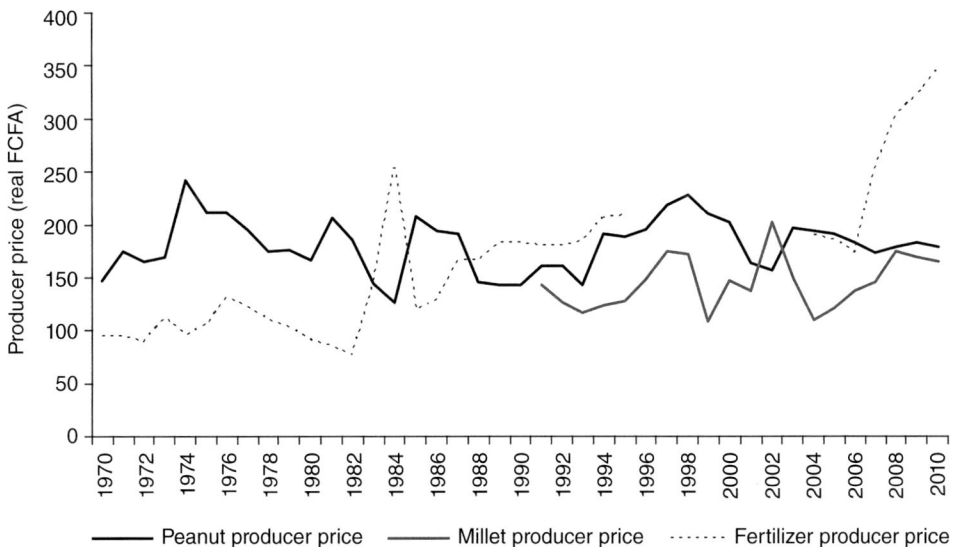

Fig. 1.4. Trends in producer prices for peanuts, millet, and fertilizer in real value (Freud *et al.*, 1997; Kelly *et al.*, 2011; FAOSTAT data).

in 2003 and 215 mm in 2014. The average length of the rainy season has recovered to 70 days over the 1999–2012 period. This extension is accompanied by the maintenance of a very high interannual variability in the rainy season's duration and the start of the first rains, for which extreme situations are becoming more frequent. In fact, from 2000 to 2010, 1 of 2 years is characterized by first rain delayed or advanced by more than 20 days compared with normal, with extreme timing in early June or the last week of August. These conditions complicate dry sowing because of the increased risk of incomplete germination and seed rot, or rodents consuming seeds before germination.

1.4.3 Dissemination of shorter-cycle varieties and adaptation of cultivated species

Climatic deterioration has led to the selection and dissemination of shorter-cycle varieties. In the central part of the peanut basin, 90-day peanut and millet varieties are replacing the 120-day varieties used before the 1970s. In Louga, in addition to the early cowpea and millet varieties (70 days) that existed before the 1970s, 80-day peanut varieties are now available. In the central and northern parts of the peanut basin, farmers are now sowing an extremely short-cycle (45-day) cowpea variety called "rescue cowpea," which can produce a crop even in years with critical rainfall.

As a result of unfavorable price trends for farmers, peanuts' place in the cropping pattern has declined since the 1960s, in favor of millet and cowpeas. As a result, peanut production is declining, while demand for peanuts is increasing on the domestic market (urban and rural). Peanuts are consumed in various forms, including peanut paste after the first pressing. It is used in the preparation of millet couscous and contributes, particularly in rural areas, to enriching the food ration with protein. This is why, in the Bambey region, the peanuts produced are now mainly intended for family consumption. The crushing plant in Diourbel has closed and been replaced by artisanal shelling and pressing activities carried out in rural towns by small traders and farmers who had the means to invest in a shelling machine and a small, motorized press.

The increased use of millet in the cropping pattern for family self-consumption, or to sell on the domestic market, illustrates the significant expansion of millet production in Senegal since 1980 in response to growing urban demand. This production has also been favored by a policy of support for artisanal, semi-industrial, and even industrial processing that allows for the milling of these cereals and their integration into urban diets, with a considerable reduction in preparation time for households (Gueye and Faye, 2010).

Although susceptible to parasitism, cowpeas, unlike peanuts, have the advantage of being easily fertilizer free. In addition, this legume with a creeping growth habit (like the peanut varieties used before the spread of animal draft cultivation) provides better soil cover and helps limit evaporation during the crop cycle. Its gradual expansion since the end of the 1980s in the central peanut basin reveals a shift of this crop toward the south. Harvested as seeds for human consumption, cowpeas can be grown in association with other crops. In Bambey, peanuts are now very often grown in association with cowpeas, and sometimes with millet, which then benefits from the nitrogen enrichment of the soil provided by this legume (Fig. 1.5). It can also be a source of fodder, as in Bambey, where all families now cultivate a field of cowpeas as a sole crop. It is harvested green, before the pods ripen, in order to build up fodder stocks to complement peanut plant residues. In Louga, cowpeas, some of which are harvested for fresh sale, now occupy more than 50% of the area cultivated on most farms.

The Senegalese government has promoted cowpeas as a substitute for peanut in the northern part of the peanut basin since the 1980s, with incentives such as increasing the official relative price of cowpeas compared with peanuts (Ndoye and Ouedraogo, 1989).

1.4.4 Intensification of labor and capital in livestock and organic matter management

1.4.4.1 *Retreat to calf production, small ruminants, and seasonal fattening*

For the past 40 years, the number of breeding animals (cows, goats, and sheep) has been

Fig. 1.5. Cowpeas associated with millet (Garambois, 2015).

declining. Only a small number of families have had the means to buy back a few cattle and now have enough fodder to feed them. Very few families have been able to reconstitute a herd that could be worth resuming transhumance, with all the risks (theft in particular) that such movements entail today. The frequency of animal theft is a further deterrent to wealthier farmers from accumulating too many livestock. Most families have only small ruminants.

On the other hand, the Bambey and Louga regions are experiencing a boom in fattening activities, which could be linked to the increased demand for meat in urban centers, particularly sheep for religious celebrations. The advantage of fattening animals is that the capital is only frozen for a short period of time, often only a few months, offering greater cash flow flexibility and a lower risk of forage shortages. This flexibility even makes the animals accessible to young working farmers and farmers with little land. The duration and intensity of fattening are directly linked to fodder resources (crop residues,

bran, locally-produced cattle oilseed cakes) and capital (purchased feed) available to families: the animals can be fattened for 3–6 months or, on the contrary, quickly sold with little creation of value-added in case of unforeseen events.

1.4.4.2 Profound changes in animal feed and organic matter management

In the face of diminishing forage resources and limited herd size, a profound change in forage management and animal husbandry has gradually taken place. In the 1970s and 1980s, farmers began to harvest part of the millet crop residues. Today, they are collected in their entirety, or almost. Millet canes are used in part to renew the fences surrounding the houses' courtyards, but now they are also used to feed draft animals, cattle, and small ruminants (breeding or fattening animals), which are increasingly kept in stalls. The animals are therefore no longer kept in common pasture and the crop residues belong to the person who cultivates the field (Photo 1.2).

Peanut plant residues, as well as millet and sorghum canes and leaves, are carefully harvested, stored, and fed to animals throughout the year, with the addition of cereal bran mush (Fig. 1.6). Supplementary feed is provided to the animals in the form of "ripass," a mixture of crushed cereals and peanut oil mill residues, or peanut cake from artisanal processing. The pruning of fodder trees at the end of the dry season allows small ruminants to be fed when the grasses and shrubs in the few uncultivated areas are no longer sufficient (Fig. 1.7).

Fig. 1.6. Stockpiling millet straw for feeding stalled animals in homes' backyards (Garambois, 2015).

Fig. 1.7. Small ruminant herd consuming the product of a sump pruning at the end of the dry season (Garambois, 2015).

Faidherbia albida has been declining over the past 40 years in the parkland associated with annual crops and is less involved in enriching the soil with nitrogen, although local farmers' testimonies point to a threefold yield under its canopy (Fig. 1.8). It is replaced by trees that are certainly fodder, but that are not leguminous trees. The testimonies collected in the Bambey region show a notable change in the parkland composition, with a general decline in the importance of *Faidherbia albida* (a rather Sudanese species), particularly in the northern part of the region. In that area, *soump* (*Balanites ægyptiaca*, a Sahelian species) is becoming increasingly dominant, whereas in the 1960s, it was mainly found in the more northerly Cayor region (Louga) (Pélissier, 1966). *Soump*'s double root system allows it to both seek water at a depth of more than 7 m and within a radius of 20 m, thanks to its deep roots, as well as take advantage of low rainfall and the slightest increase in soil moisture, thanks to a superficial root system located at a depth of 2 or 3 cm.

Faidherbia albida is nowadays less favored in the parkland associated with crops because its seeds' germination is facilitated by their prior ingestion by ruminants. However, the management of ruminants has changed considerably. As they are increasingly kept in stalls, animals have much less opportunity to ingest these seeds. In this context, the labor intensification in feeding herds and manure management aims to compensate for the reduced availability of manure by increasing the efficiency of restitutions on cultivated fields. Some farmers even go so far as to collect *Faidherbia albida*

pods in the field, which are distributed to small ruminants that are kept for fattening near the houses. This contributes to enriching their diet, but more importantly spreading their excrement on cultivated fields, and with it seeds that can germinate. This effort aims to ensure the digestion stage, which is essential for maintaining *Faidherbia albida* in the cultivated plots.

Finally, farmers who have enough fodder leave a small amount of crop residue spread on the field during the dry season. The purpose of this seems to be to try to limit evaporation and wind erosion during the Harmattan winds. The straw is raked at the end of the dry season (Fig. 1.9) and burned. The ashes are then scattered on the field in preparation for sowing.

1.5 Major Inequalities in Adaptation and Sensitivity to Hazards

While most rural households in the Louga and Bambey regions have found ways to adapt to the profound climatic and economic changes of recent decades, not all are equally placed in this process. The peanut basin provides an illuminating example of inequalities in adaptation and vulnerability to hazards.

1.5.1 Off-farm income and migration, deep inequalities in access to capital

The agricultural income levels per farmer recorded by the various categories of producers in central

Fig. 1.8. Progression of *Balanites aegyptia* in the wooded parkland (north of Bambey) and effect of *Faidherbia albida* on the growth of crop species under its canopy (Garambois, 2015).

Fig. 1.9. Raking millet straw left in the field throughout the dry season to limit evaporation and soil erosion (Garambois, 2015).

and northern Senegal are often low, or even extremely low, and sometimes far from allowing these families to survive without additional external income (see below). Many families have one or more members working in the city, either permanently or temporarily. Their activity type is often strongly correlated with their families' resources in the village. Sometimes these are qualified and permanent jobs that allow educated young people to send regular and substantial amounts of money to their families for productive investment.

In the Bambey region, migration is most often seasonal and tends to be toward Thiès or Dakar. Activities are quite diverse: bus driver assistant, small informal street trade, porterage, domestic work, and for some, transport with carts purchased with the capital accumulated through farming. Above all, the salaries paid cover the costs of food and basic housing for these young people. The meager amount left over allows the family in the village to buy a little

rice or it can be used as security for recourse to buy a small ruminant to fatten.

This migration phenomenon is not new, but it is more recent in the central peanut basin. Roquet (2008) also mentions that seasonal migrations are much more numerous when the previous wintering rainfall was below average, and the production deficit is significant. Although the drought did not trigger the migration phenomenon, it does seem to intensify its scope.

In the rural area studied south of Louga, only the households of the children after the first-born child, underequipped and relying solely on themselves to support their families, remain in the village year-round. They conduct some supplementary handicraft activities and hire out their hands for agricultural tasks. All other families now have at least two members (son or younger brother) working in the city: in Dakar, Thiès, or Louga. Most of them manage to return to the village occasionally to help with agricultural work during peak labor periods. Their wives and children remain in the village. In most rural households, there is a clear imbalance between men and women outside the rainy season.

These migratory phenomena and the importance of off-farm income for families' survival and productive investments in the village are older and more extensive in the north of the peanut basin. In this region, labor migration increased sharply in the 1970s, due to the dramatic drop in harvests (Ndoye and Ouedraogo, 1989). This migration toward the large Senegalese cities was sometimes only a first step toward expatriation. The Louga region is indeed the one where the share of migration to Europe was the highest in all of Senegal (more than 75%), as indicated by the 1993 and 2002 national surveys (Lessault and Flahaux, 2013). These early (from the end of the 1980s) and distant departures mostly applied to the richest families in the region, who were able to mobilize the necessary capital to send one of their members to Europe in the middle of a drought. This was a real investment for the family, thanks to the money sent back from abroad. Comparatively less expensive at the time, the cost of these trips to Europe for the family is now close to five million CFA francs (about 7500 euros). With one family member emigrating to Europe and two or three others working in Dakar, the wealthiest families are also those with the fewest men involved in agricultural activities. This deficit is compensated by the employment of one or two salaried employees.

1.5.2   Inequalities in access to fertility, animal draft power, and land

With the end of free-range common grazing, the amount of animal dung available to a family is now correlated to the crop residues available on their fields (and thus the cultivated area), and their capacity to accumulate livestock and purchase supplements. In Bambey, the poorest families are sometimes reduced to replacing part of the feed purchased for their fattening sheep with mush prepared from shredded cardboard.

This unequal access to livestock and manure is coupled with unequal access to chemical fertilizer. Although the Senegalese government has been organizing the distribution of a few bags of fertilizer to each village, subsidized at a rate of 50% since the end of the 2000s, these bags rarely arrive on time and in quantities too small to meet the entire local demand. The purchase of selected peanut seeds and fertilizer for peanuts or millet is therefore now limited to the wealthiest families, who have sufficient cash to purchase non-subsidized inputs. In the northern part of the peanut basin, the high degree of specialization in peanut production (about two-thirds of the area sowed) by families with expatriate members is directly linked to their ability to purchase chemical fertilizer and selected peanut seeds thanks to their migratory income.

The unequal access to mineral elements from one category of producer to another is thus maintained, and even accentuated, as the price of fertilizer evolves unfavorably for producers and as manure becomes rarer and all the more precious. In Bambey, for example, there are considerable differences between two types of families: those who have had the means to reconstitute cattle and small ruminant herds and have been able to fatten them; and the poorest families, who have at best only two or three goats or sheep (or even no animals at all). The latter are reduced to collecting a few droppings still available on common pasture (residual fallow land and roadsides) in addition to the organic fertilizer produced by composting kitchen waste.

Very different quantities of manure can be applied to fields, depending on its availability, the capacity to transport it, and the fields' distance; the furthest away from the village being less manured overall. While some families are able to spread animal dung abundantly on their fields, others have to make do with smaller volumes of dung supplemented with organic fertilizer from kitchen waste, while others have to make do with burning only the meager straw residues in the field and kitchen waste.

Access to animal draft power and the various equipment linked to it (cultivator, seed drill, lifter, cart) is also extremely variable. It determines the cultivable area per farmer and even the type of crops that can be grown, especially those that are sown after the first rain. In the Louga region, the rainy season window is particularly short. Peanut sowing just after the first rainfall must be done within 3–5 days, never more than a week. This operation requires 1 day's work per hectare in donkey draft power, but almost half that in equine draft power. A distinction is made between:

- Families with equine draft power and all their equipment (and as many horses, cultivators, and seed drills as there are men of working age) who can cultivate 1.2 ha per farmer
- Families with only donkey draft power and no lifter who cultivate 0.6–0.8 ha per farmer
- Families without access to animal draft power; they occasionally borrow equipment (0.4 ha per farmer) or work manually (less than 0.2 ha per farmer) and do not have the means to produce peanuts

In force since 1964, the law on national domain does not recognize or officially regulate indirect tenure. Although rarely acknowledged because it is illegal, tenant farming is nevertheless common in the region. For example, older farmers, whose children have moved to the city to take up more lucrative jobs, may rent out part of their land. Lessees may be long-established families in the village with sufficient family labor and the level of equipment required to cultivate larger areas. They may also be among the poorest or most recently arrived, in a very precarious position. The law provides that a field not cultivated by its owner for more than two consecutive years must be transferred to the national domain. This means that owners do not risk renting the same field to the same farmer for more than 1 or 2 years, and lessees are less likely to try to increase the fertility of the soil on them.

1.5.3   Inequalities in self-sufficiency and food security

In the Bambey region, millet now plays a dominant role in the cropping pattern, and its importance tends to increase as the area of land per farmer available to families decreases. Most families cultivate millet on 75–80% of their land, while the rest is sown with a combination of peanuts and cowpeas strictly for family consumption. There are large differences in yields from one production system to the other. Families with less than 0.4–0.5 ha per farmer are equipped with a cultivator and a cart but must borrow the seed drill. In addition to their pack animal, these families own a few small ruminants

or take on a cow for herding. They manage to have some manure and to achieve millet yields of 400–450 kg/ha, sufficient to provide the family with a daily millet meal for 7–9 months of the year. The rest of the meals depend on the purchase of rice. Families who are even poorer, with the same equipment but only 0.3–0.4 ha per farmer and no breeding animals, have lower yields (350 kg/ha) and can only provide a daily meal of millet for 4–5 months of the year. In contrast, families with 0.7–0.9 ha per farmer, full animal draft power, more animals (one cow and one sheep per hectare, plus some for fattening), and thus more manure, can provide a daily millet meal to their families all year-round.

The richest families, who cultivate more than 1 ha per farmer, devote 60–65% of their land to millet. The rest is devoted to peanuts associated with cowpeas. These families have all the equipment needed for animal draft power and a "lifter" for digging up peanuts. They employ women from the poorest families to harvest peanuts. In addition to guaranteed year-round food self-sufficiency for the evening millet couscous, they are able to generate surplus peanuts and millet. These families can buy fertilizer and have large manure quantities, thanks to their sheep and cattle herds. They obtain millet yields of nearly 1 ton per hectare. Some of them are extended families made up of several nuclear households, each with a complete set of animal draft power equipment. The size of the reconstituted herd at the extended family level then justifies resuming transhumance in the rainy season.

In the Louga region, millet cultivation plays a much smaller role in the cropping pattern. More importantly, this role diminishes rapidly as off-farm and migratory income increases within the households' total income. Thus, millet accounts for only 5–10% of the land cultivated by families with one member in Europe and only one-third of the male labor force engaged in agriculture. This share rises to 20–25% when 50% of the male family labor force is engaged in off-farm activities. It reaches 30–50% in the poorest nuclear families, which have no (or rare) access to animal draft power, having little off-farm income and therefore prioritizing family consumption. Conversely, families with high off-farm incomes can specialize more extensively in the production of pulses, especially cowpeas, and purchasing the bulk of the cereals consumed by the family (millet and rice).

1.5.4 Uneven development of higher value-added crops and processing activities

Cultivated today by all families, bissap has developed rapidly over the last 10 years. This labor-intensive crop (manual harvesting of the flowers) has the advantage of being very profitable and inexpensive in terms of inputs: the seeds can easily be saved from one year to the next and no fertilizer is required. The white variety, used as a condiment in millet couscous or rice dishes, is generally reserved for family consumption, while the red variety, used in the preparation of drinks, is sold. Sown at the edge of fields so as not to compete with other crops, the volume of bissap produced is directly related to the area per farmer available to families. In the Bambey region, the poorest families market 20–30% of their production (2 basins per year), while the richest families can sell up to 70% of their production (30 basins per year). Fresh cowpeas, harvested green as early as 30 days after sowing, can also be sold.

Watermelon is the only vegetable crop not requiring irrigation that has developed in recent years in the Bambey region. Buyers/retailers come to the villages to buy the entire standing crop. For most producers, sowing is done in late September and harvesting takes place in December–January, without requiring any phytosanitary treatment. Only farmers with sufficient cash flow can undertake the production of "early" watermelon (sowing at the end of July and harvesting in early October). It proves to be more profitable, but at the cost of numerous pesticide applications during this season: one insecticide application per week for nearly 3 months. The treatments are then carried out with a small tree branch soaked in the treatment solution that is shaken over the plant. Today, the richest and most landed families can devote up to 15% of their area to watermelon cultivation and have equipped themselves with a backpack sprayer to ensure better effectiveness of the treatments.

Thanks to their off-farm income, many of the wealthier families in the central and northern parts of the peanut basin have invested in peanut shellers and jack presses. This equipment allows them to add value to their peanut seeds by going to the final stage of artisanal processing: oil and oilcake (Fig. 1.10), while at the same time amortizing the investment by providing services to other local farmers. The trade and artisanal processing of peanuts has developed in parallel with the sector's gradual liberalization, decline in peanut yields, unfavorable evolution of relative prices, and decline in value-added recorded at the field level. They complicate the supply of industrial oil mills because they offer greater flexibility to producers by allowing them to market their peanuts when they want to and outside the official oil mill purchasing season. In a context of declining cultivated peanut area, factories are struggling to obtain supplies and make their industrial tools, which are sized for much larger volumes, profitable (Gaye, 2010). This explains why this artisanal processing method, which diverts part of the peanut seeds from the factories, is prohibited, the official argument being sanitary reasons.

1.5.5 Wide income disparities between farm households

1.5.5.1 Deep inequalities in farm income in the central peanut basin

The differences in farm income recorded in these regions today—here presented including production for family consumption—reflect families' unequal access to means of production. This access affects labor productivity, the ability to generate marketable food surpluses, and the ability to engage in higher value-added productions.

In Bambey, the poorest families have very little cultivable land per farmer, on land where they are sometimes only precarious tenants. They are underequipped compared to other categories of producers. With little or no livestock, they can only ensure very low returns of fertility. These families only manage to ensure about a third of their cereal needs and have extremely low incomes per farmer, below 50,000 CFA francs (75 euros). They may be young households working partly outside the home or elderly households.

Families with medium-sized areas (0.4–0.6 ha per farmer) manage to generate 75,000–150,000

Fig. 1.10. Artisanal peanut processing and products from its processing (oil and oilcake) (Le Goff, 2016).

CFA francs (110–220 euros) per farmer by developing higher value-added production (watermelon), in addition to food crops and their few small ruminants (fattening). Families that can work 0.6–0.8 ha per farmer and have reconstituted their herd or engaged in even higher value-added production (watermelon and fattening) have a farm income of 100,000–200,000 FCFA (150–300 euros) per farmer. Finally, the categories of producers with the best equipment, capital (cash and livestock), and land (0.8–1.1 ha per farmer) manage to generate an income of 200,000–300,000 FCFA (300–450 euros) per farmer.

1.5.5.2 Inequalities reinforced by labor migration in the northern peanut basin

In the region studied, the northern part of the peanut basin, inequalities in farm income are also marked by a factor of one to six. The best-off households live in extended families. They may be families that cultivate large areas of peanuts (PS 1.1, PS 1.2, PS 1.3; Fig. 1.11) thanks to their larger lands, equipment, and livestock, or they may be herder families that have been able to reconstitute large transhumant herds (PS 4; Fig. 1.11). At the opposite end of the social scale, some young nuclear households, with little land and only a few small ruminants, practice manual farming (PS 3.2; Fig. 1.11). There is still a one-to-three ratio between these precarious farmers and those with animal draft power but less cash than the better-off families (PS 2.1, PS 2.2, PS 2.3; Fig. 1.11).

These income differences are amplified by off-farm income—measured here after deducting housing and transportation costs in town. Families with an expatriate member receive approximately 1,000,000 CFA francs (1500 euros) per year. The differences can thus reach a factor of one to nine between families with one member in Europe and two or three others working full time in the city, and nuclear households that derive their off-farm income from agricultural work in the country.

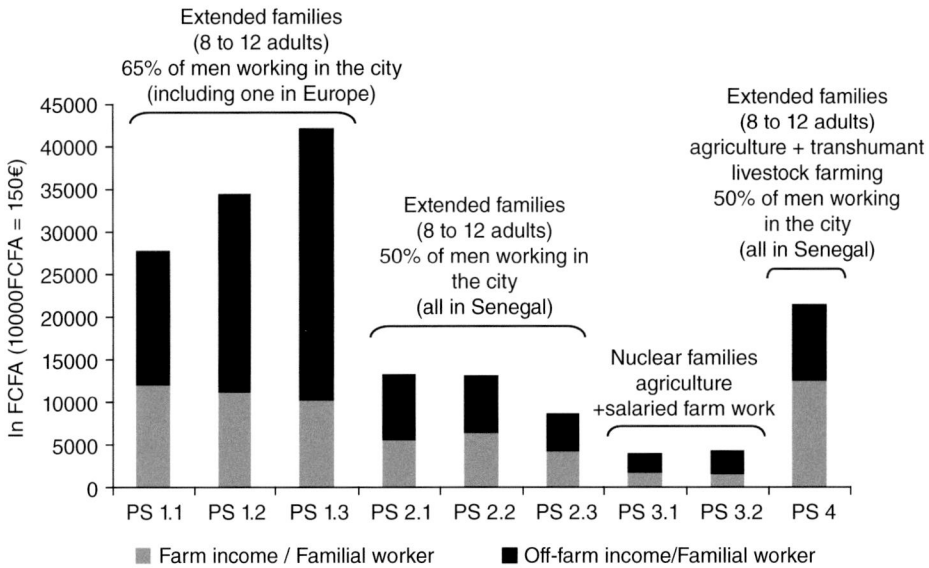

Fig. 1.11. Income level and composition per family farmer by production systems in the rural area south of Louga (Le Goff, 2016).

1.6 Vegetable Crops, Adaptation Through Capital and Labor Intensification Strategies?

The southern part of the Bambey region benefits from a semi-profound water table with a salt content suitable for agricultural use. In response to drought, international solidarity associations initiated the establishment of small village irrigation perimeters some 30 years ago (Fig. 1.12). Boreholes of about 80 m deep were dug. Irrigated perimeters were installed on some villagers' fields after negotiation and retrocession of a field within the irrigated perimeter. The allocation of the remaining fields was usually based on a consultation organized by the village authorities, with preference given to young producers who were deemed capable of growing approximately 1200 m² of vegetable crops each (an area that occupies a full-time family farmer).

In addition to financing all the necessary heavy investments (drilling, buried pipes, and individual tanks equipped with a water meter), these associations distributed a range of tools to each producer cultivating in the irrigated perimeter (wheelbarrow, rake, hoe, watering cans, shovel, dibble, pitchfork, and rope), equipment that has since been replaced. They also organized

3-month technical training courses and provided a cash fund to act as an annual farming credit.

Vegetable crop production is organized in two seasons. The first one takes place during the rainy season, with an identical production plan for all producers (1200 m² per producer divided into 45 cabbage beds, 20 tomato beds, 5 pepper beds, 5 eggplant beds). Production is carried out in groups, controlled by an office of the producers' group in charge of buying inputs and negotiating the prices of vegetable products with intermediaries ("bana-bana") who buy the production to resell it in Dakar. The second season focuses on onion production, which is cultivated more freely by each farmer, in terms of both technique (which is dependent on their ability to provide manure) and marketing (each producer is responsible for selling their production to traders or locally).

Supported by NGOs for about 10 years, these perimeters have since been operating at cruising speed without their support. They seem to enable farmers to obtain much higher incomes per surface area unit, but at the price of labor-intensive agricultural tasks, great vigilance in terms of cash flow, and know-how patiently developed over the past 30 years. For the time being, farmers in the few existing perimeters are

Fig. 1.12. Vegetable crops in collective irrigated perimeters at the end of the dry season in the Bambey region (Garambois, 2015).

managing to find outlets for their production at prices that remain remunerative. However, these producers in the peanut basin carry little weight compared to producers in the Niayes region, which is largely specialized in vegetable cultivation. On the other hand, they do have the comparative advantage of a later vegetable crop production period compared with the main vegetable crop production areas in Senegal.

The Louga region is currently experiencing only a timid development of off-season vegetable crops. Attempts were made along the Lake Guiers pipeline but were quickly banned. In recent years, a few rare private initiatives have emerged among families with sufficient capital (with a relative in Europe) to invest in a well and equip themselves with a pump. These small vegetable crops areas (Fig. 1.13, Photo 1.3), run with hired labor, are irrigated using the first and shallowest water table (35 m), previously reserved for human and animal drinking water. The investments required to tap into the water table located 80 m deep remain inaccessible to individuals.

1.7   Conclusion

Strictly manual until the 1960s, rainfed agriculture in the north and center of the Senegalese Sahelian Peanut Basin has long been able to adapt to difficult and uncertain rainfall conditions. To do so, it has been able to rely on the choice of short-cycle species and varieties, herd mobility, gathering, and off-farm income (processing, handicrafts, and small-scale trading). In half a century, it has undergone profound upheavals: the agricultural revolution of the 1960s, the profound rainfall declines of the 1970s–1990s (drop in yields and available fodder, acute decapitalization), and the almost total cessation of public support for peanuts from the early 1980s. However, rural households have been able to adapt by:

- deeply modifying the link between crops and livestock, as well as herd husbandry, thanks to labor intensification (collection of crop residues, animals in stalls, transport of excrement on the fields)
- changing their cropping pattern in favor of millet and cowpeas, which are better adapted to rainfall hazards and less cash-intensive than peanuts
- making greater use of early, or even very early, varieties developed by Senegalese agronomic research

For the peanut basin, these trends seem to be confirmed since the mid-2000s (Dieng and

Fig. 1.13. Offseason vegetable crops on private land in the northern peanut basin (Le Goff, 2016).

Gueye, 2005). Nevertheless, many rural households can only ensure very low returns of fertility to their cultivated fields (organic matter, fertilizing elements), while the place of *Faidherbia albida*, a true "nitrogen pump," continues to decline within the parkland associated with annual crops. This phenomenon suggests that the fertility crisis in these production systems is worsening.

In Bambey, rural households still mainly live from agriculture. The beginnings of more intensive fertility transfers, which would both increase yields and promote the densification of the *Faidherbia albida* parkland, seem to be in the making among families. Although they are not the best endowed with land or livestock, they nevertheless have mechanized equipment. In the absence of grazable fallow land, these families are striving to make forage cowpeas more important in their cropping pattern, in order to increase the volume of fodder and animal manure collected at the stall. By concentrating the manure's use on millet and peanut areas that are *de facto* smaller, fertility transfers are amplified, which contributes to increased yields, both in grain and crop residues.

This remarkable capacity to adapt, as emphasized by Jouve (1991) for the Sahel as a whole, and the maintenance of rainfed agriculture under these conditions, which, despite a certain recovery in rainfall, remain very uncertain, should not make us forget the profound inequalities that remain and are increasing among rural households, as well as the central role that off-farm income has played and still plays. Today, some families do have a complete set of animal draft power equipment, large areas under direct tenure, and relatively abundant manure thanks to a progressive recapitalization in livestock, and they can purchase fertilizer. They thus generate marketable surpluses and, particularly in the north, can count on substantial migratory income. By default, the most precarious households practice manual farming or are dependent on borrowing animal draft power equipment. They have very little land at their disposal, even if they have recourse to indirect tenure, which is informal and precarious. These households are reduced to hiring out their hands in the village or sending their children to the city in the dry season to carry out low-skilled and low-paid activities.

This unequal ability to adapt and the specific antirisk techniques used by rural households, depending on their means of production, highlight the central role that agricultural and land policies can play in promoting these agricultures' adaptation and limiting their vulnerability to the dual hazards of climate and prices.

References

Chalby, N. and Demarly, Y. (1991) *L'amélioration des plantes pour l'adaptation aux milieux arides*. John Libbey Eurotext, Paris, France, 228 pp.

Chevalier, A. (1929) *Rapport sur les moyens d'intensifier et d'améliorer la culture de l'arachide au Sénégal*. Dakar, Senegal, 95 pp.

Chevalier, A. (1947) Amélioration et extension de la culture des arachides au Sénégal. *Revue internationale de botanique appliquée et d'agriculture tropicale* 295–296, 173–193.

Debien, M.T. (1966) L'association du Sénégal à la communauté économique européenne et les problèmes de l'arachide. *Bulletin de l'Ifan* 28B(3–4), 878–926.

Dieng, A. and Gueye, A. (2005) *Revue des politiques agricoles au Sénégal: bilan critique de quarante années de politique céréalière*. 25 pp. Available at: https://www.bameinfopol.info/IMG/pdf/Revue_Dieng_et_Gueye.pdf (accessed 7 August 2024).

Freud, C., Hanak Freud, E., Richard, J. and Thenevin, P. (1997) *L'arachide au Sénégal: un moteur en panne*. Khartala-Cirad, Paris, France, 166 pp.

Gaye, M. (2010) L'arachide en crise: baisse des prix, retrait de l'Etat et concurrence sur le marché des huiles. In: Duteurtre, G., Faye, M.D. and Dieye P.N. (eds) *L'agriculture sénégalaise à l'épreuve du marché*. Karthala/ISRA, Paris, France, pp. 113–136.

Gueye, A. and Faye, M.D. (2010) Mils et sorghos: l'émergence d'un artisanat agro-alimentaire. In: Duteurtre, G., Faye, M.D. and Dieye, P.N. (eds) *L'agriculture sénégalaise à l'épreuve du marché*. Karthala/ISRA, Paris, France, pp. 83–96.

Jouve, P. (1991) Sécheresse au Sahel et stratégies paysannes. *Sécheresse* 2, 61–69.

Kelly, A. (1989) Vers une meilleure connaissance de la demande pour les engrais dans le bassin arachidier. *Actes du séminaire: La politique agricole au Sénégal, ISRA* 1(2), 6–23.

Kelly, V., Faye, M., Ndiaye, J.-P., Fall A.A. (2011) *Projet 'croissance économique': Analyse de la filière engrais au Sénégal et de son évolution sur la période 2000 à 2010. Rapport de synthèse*. Agence Américaine pour le Développement international (USAID), Washington, Etats-Unis, 16 pp.

Le Borgne, J. (1988) *La pluviométrie au Sénégal et en Gambie*. Laboratoire de Climatologie, Université Cheikh Anta Diop, Dakar, Senegal, 28 pp.

Lessault, D. and Flahaux, M.-L. (2013) Regards statistiques sur l'histoire de l'émigration internationale au Sénégal. *Revue européenne des migrations internationales* 29(4), 59–88.

Ministère de la coopération et du développement. Sous-direction des études du développement. (1983) *Marché mondial des oléagineux: perspectives pour les pays africains producteurs d'arachide*. Ministère de la coopération et du développement, Paris, France, 130 pp.

Minvielle, J.-P. and Lailler, A. (2005) *Les politiques de sécurité alimentaire au Sénégal depuis l'indépendance*. L'Harmattan, Paris, France, 187 pp.

Ndoye, O. and Ouedraogo, I. (1989) Commercialisation des produits agricoles dans le bassin arachidier: Situation actuelle et implications pour la politique agricole. Actes du séminaire: La politique agricole au Sénégal. *ISRA* 1(2), 128–140.

Oya, C. (2009) Libéralisation de la filière arachide. In: Dahou, T. (éd.) *Libéralisation et politique agricole au Sénégal*. Crepos—Karthala—Enda/Graf/Diapol, Paris, France, pp. 97–125.

Pélissier, P. (1966) *Les paysans du Sénégal: Les civilisations agraires du Cayor à la Casamance*. Fabrègue, Paris, France, 939 pp.

Portères, R. (1950) L'assolement dans les terres à arachide du Sénégal. *Revue internationale de botanique appliquée et d'agriculture tropicale* 327–328, 44–50.

Roquet, D. (2008) Partir pour mieux durer: la migration comme réponse à la sécheresse au Sénégal? *Espaces, populations, sociétés* 2008(1), 37–53.

Salack, S., Muller, B., Gaye, A.T., Hourdin, F. and Cisse, N. (2012) Analyse multi-échelles des pauses pluviométriques au Niger et au Sénégal. *Sécheresse* 23, 3–13.

Sambou, P.C., Sagna, P. and Yade, M. (2015) Évolution climatique récente, productions agricoles et stratégies d'adaptation des paysans dans les communautés rurales de Mbediene et de Leona (Département de Louga). *Revue de géographie du laboratoire Leïdi* 13, 117–131.

USAID. (2011) *Projet 'croissance économique': Analyse de la filière engrais au Sénégal et de son évolution sur la période 2000 à 2010*. Rapport de synthèse, 16 pp.

2 In the Highlands of Eastern and Southern Africa: A Fragile Rainfed Agriculture, But Supported by Small-Scale Irrigation

Hubert Cochet[1]*, Louis Thomazo[2], Esther Laske[3], and Niel Verhoog[4]

[1]*Comparative Agriculture Training & Research Unit/UMR Prodig, AgroParisTech, Palaiseau, France;* [2]*Agroeconomiste Roma, Italy;* [3]*The World Bank, Washington D.C.;* [4]*CEZ-Bergerie Nationale de Rambouillet, Rambouillet, France*

Three agricultural areas that are predominantly rainfed, but where small-scale irrigation is also essential, were studied in detail in this research program. Two of them are in Zambia, the Katongo Kapala site in Mpika District (1500 m altitude) and the Miloso site in Mkushi District (1400 m).[1] The third region is located not far from Iringa, in the southern Tanzanian highlands (at an altitude of 1800 m, Fig. 2.1).

In all three regions, the main food (and cash) crop, corn, is suffering from the shortening of the rainy season and is increasingly perceived as a "risky" crop. Off-season crops are also widely grown, either in the lowlands based on residual moisture or through small-scale irrigation schemes planned by farmers below springs or by diverting streams with small elevator dams and feeder canals.

The detailed study of these three regions therefore allows us to address both the vulnerability of rainfed agriculture in this part of the world and that of small-scale irrigation as a possible way to develop this agriculture, making it less subject to hazards while anticipating the potential climatic deterioration, which has been forecasted for the long term.

2.1 Climate Change: Is Rainfed Agriculture Threatened?

2.1.1 A short but relatively generous rainy season

In the northern highlands of Zambia, the rainy season is relatively generous (900–1100 mm of rain) but lasts only 4–5 months. The rains normally start in November, intensify in December, January, February, and March (150–200 mm/month), and then stop quite abruptly in the first days of April (Fig. 2.2). The Mkushi and Mpika regions, which will be discussed in this chapter, are therefore not among those in Zambia usually cited as being the most vulnerable in terms of exposure to drought or flooding risks. Here, it is the interannual rainfall variability that is very marked and clearly mentioned by producers as

*Corresponding author: hubert.cochet@agroparistech.fr

DOI: 10.1079/9781800628137.0002

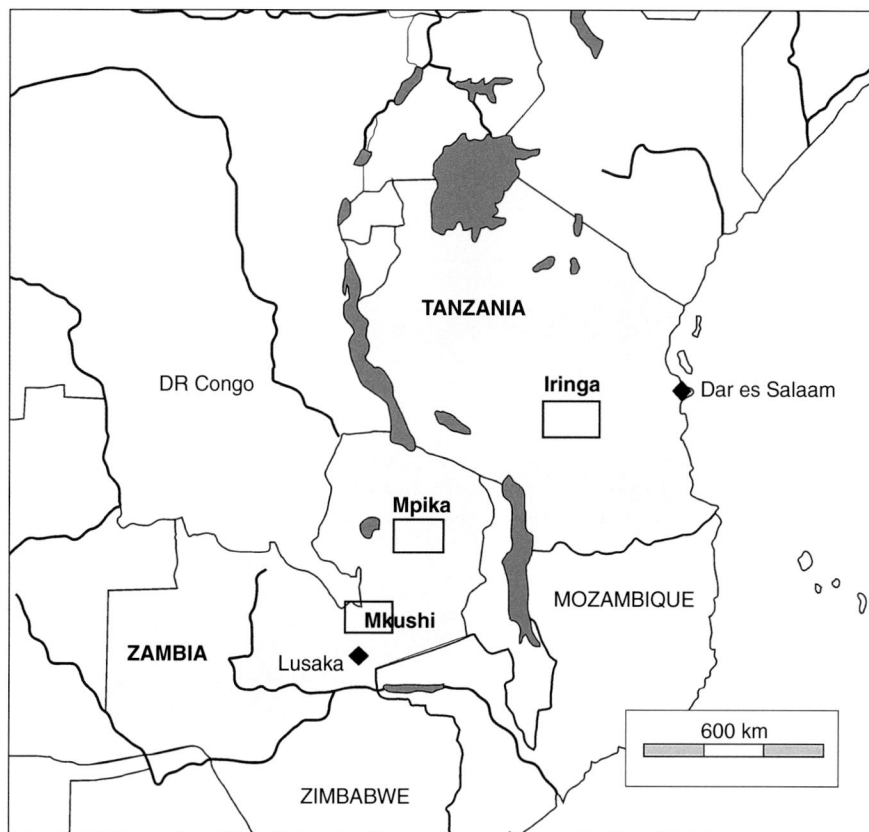

Fig. 2.1. Location of the three study regions in the Tanzanian and Zambian highlands.

the main hazard. In Iringa, Tanzania, the rainy season lasts a little longer from mid-November to mid-April, but average rainfall volumes are much lower, more in the order of 600–700 mm (Fig. 2.2).

2.1.2 IPCC forecasts for Southern and Eastern Africa

2.1.2.1 At the regional level

In terms of temperature trends, the fifth IPCC report (IPCC, 2014) leaves little doubt that they are likely to increase, following the trends observed over the past few decades (IPCC, 2014, p. 1206).

With regard to the possible evolution of precipitation, forecasts are uncertain and do not allow for a very clear trend. While decreases in rainfall are very likely in South Africa (as in North Africa), as well as increases in rainfall

envisaged for parts of Central and Eastern Africa (IPCC, 2014, p. 1210), the situation for the regions between the southern half of Southern Africa on one hand and the Great Lakes region of Africa on the other is much less clear; forecasts differ from model to model. Significant changes could occur in the seasonal distribution of rainfall, with more intense rainfall at certain times of the rain cycle (IPCC, 2014, pp. 1209–1210). However, all models seem to agree on the likely increase in the frequency and intensity of extreme events.

2.1.2.2 In Zambia

The literature on Zambia specifically has evolved in recent years as models and results from IPCC experts have evolved. A few years ago, temperature (increasing) and precipitation (decreasing) were projected to change together across the country (Mulenga Bwalya, 2010, p. 6). The latest studies appear to be more accurate and nuanced.

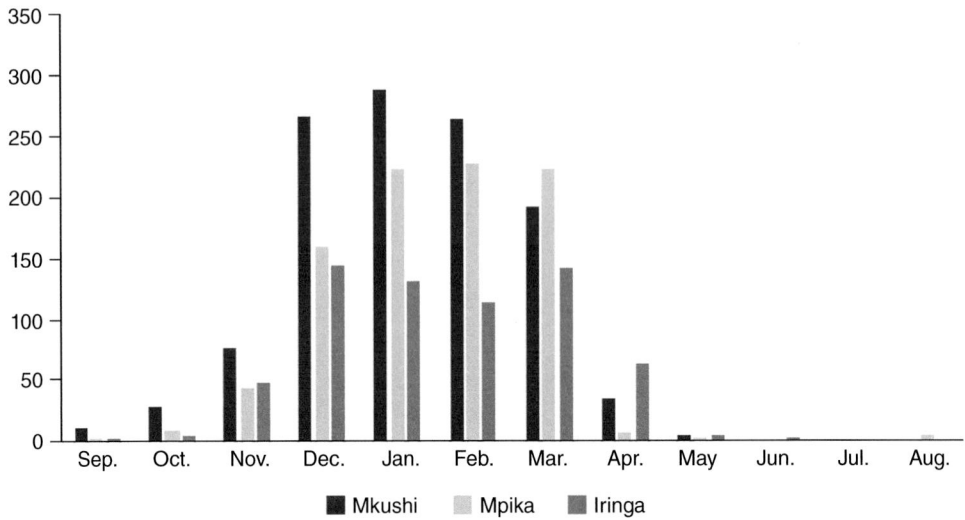

Fig. 2.2. Rainfall distribution in the three regions studied (Mkushi and Mpika in Northern Zambia, Iringa in the south-western Tanzanian highlands). (From Mkushi: Laske, 2014 based on Z&A 2013, average 2000–2010; Mpika: Thomazo (2014) and Mpika weather station, average 1993–2012; Iringa: Verhoog, 2015 and Iringa weather station, average 2000–2014.)

The consensus is growing regarding increasing temperatures, with models differing only in expected changes' magnitude (Kanyanga *et al.*, 2013).

On the other hand, the four general circulation models (GCMs) proposed for precipitation evolution deliver different results, notably with an increase in precipitation in some regions (Kanyanga *et al.*, 2013).

Predictions of increased extreme events remain convergent in all models cited (Mulenga Bwalya, 2010, p. 6).

2.1.2.3   In the south of Tanzania

Although southern Tanzania is among the regions for which an increase in rainfall is considered likely, here too models are far from converging on the magnitude of expected changes (Kilembe *et al.*, 2013). Moreover, despite the fact that all models suggest an increase in rainfall in the southern half of the country, we will see that recent rainfall trends in the regions studied in this research program do not necessarily suggest this. Moreover, southern Tanzania is indeed one of the regions that would be affected—all models seem to converge on this point—by an increase in the frequency and intensity of extreme events, especially heavy rains.

2.1.3   About the forecasted evolution of corn yields

As corn is the main cereal crop in these two countries and given its considerable importance in household consumption,[2] the work on adaptation to climate change conducted within the framework of the International Food Policy Research Institute (IFPRI) gives pride of place to models predicting changes in corn yields. Based on predictions of increased temperature and rainfall for this region of the world (*supra*), Kanyanga *et al.* (2013) envision improved conditions for rainfed corn production in Zambia, and thus increased yields, particularly in the regions where the two territories studied in this chapter are located (Mpika and Mkushi). For the Iringa Highlands in southwestern Tanzania, the arguments put forward by IFPRI experts point in the same direction—a likely increase in yields—although one in four models leads to a diametrically opposed conclusion (Kilembe *et al.*, 2013, p. 332). In the face of such discrepancies among the models, the authors cautiously conclude:

> "The results do suggest that it would be helpful to prepare a number of different responses from which farmers could choose as appropriate. It is also helpful to note that although there can be

negative consequences of climate change in some places, there may be positive outcomes in other places [. . .]" (Kilembe *et al.*, 2013, p. 332).

Insofar as public policies implemented in the name of adaptation in these two countries reserve a large place for corn and the development of technical packages designed to increase its yield, such caution with regard to this type of model's results seems appropriate,[3] especially since the latter are often invoked as scientific proof by the policies in question. However, the proposed models sometimes go much further: once the link between climate change and changes in yields has been established, the aim is to deduce future changes in the prices of the commodities in question and then to model the impact of these price changes on household poverty. A direct link is thus established between climate change and poverty, one that is presented as obvious and direct, even though a large number of factors also play a role in this complex relationship (Leichenko and Silva, 2014; IPCC, 2014).

One such model is for Tanzania (Ahmed *et al.*, 2011). The stated objective is to examine the link between climate change, changes in corn yields, and poverty, using a quantitative and modeling approach, based on both past and future trends, over a time step of 60 years (1971–2031). The demonstration is based both on the use of the various climate models available and on the

general equilibrium economic model. For the authors, the conclusion is clear: it is climate change that explains poverty through the decline in grain yields. Only policies aimed at increasing corn yields—in particular through the technical packages discussed below—and in a context of free competition, are likely to allow for a virtuous "adaptation" to climate change. We will return to this type of reasoning at the end of the book (Chapter 8, this volume).

2.1.4 What is climate change on the local scale?

2.1.4.1 The example of the Mpika region in the Zambian highlands

In the case of the Mpika region (Northern Province), Louis Thomazo analyzed the data collected locally by the Mpika meteorological station and attempted to identify local trends in climate change over the last 20 years. With regard to the evolution of precipitation, the impression that emerges from reading the data (Fig. 2.3) is the strong interannual variability. Furthermore, "calculation of a linear regression for this curve yields a positively sloped line (slope of 7.4 mm/year) indicating a slight upward trend in precipitation over the past 20 years" (Thomazo, 2014).[4]

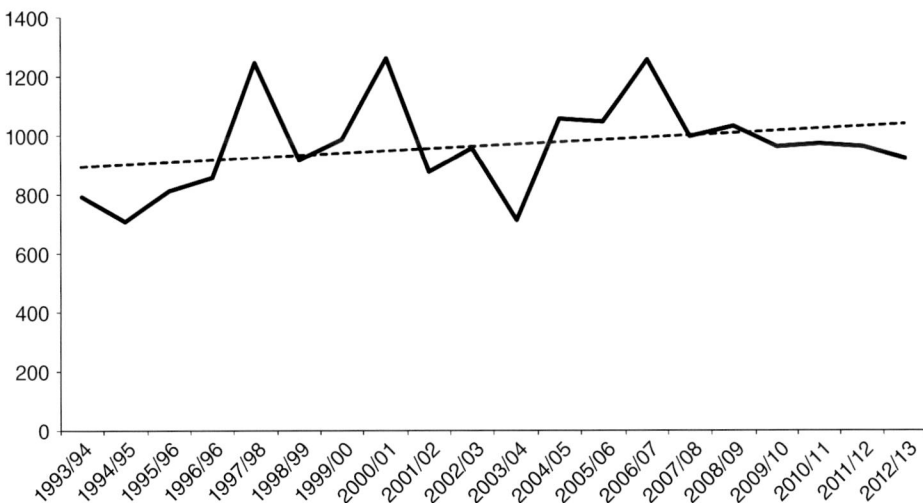

Fig. 2.3. Changes in annual rainfall at Mpika (1993–2012). (From Thomazo, 2014, based on data collected at the Mpika weather station.)

Furthermore, the study of the curve showing the amount of rainfall on the rainiest day of each year over 10 years (Fig. 2.4) lends credence to the thesis that there has been a clear increase in the intensity of violent events, although there is not yet enough time to interpret the phenomenon.

With respect to temperature, data provided by the Mpika City weather station allow the reconstruction of the month-by-month temperature trends for the past 20 years. No upward or downward trend is evident in the average annual temperature (Thomazo, 2014). Analysis of the monthly data reveals that of the four warmest months of the year, three of them (July, September, and October) have seen an increase in temperature over the past 20 years. The relative stability of average temperatures would then conceal a tendency for the cold months to cool and the warm months to warm. Furthermore, given that the temperature always cools at the end of the dry season, thanks to the arrival of clouds announcing the first rains, the perceptible increase in temperature in October could indicate a delay in the cloud's arrival date and therefore a delay in the season's first rain, a trend that is not yet detectable on the rainfall curves, but that farmers frequently mention (Thomazo, 2014).

The most obvious observation is the strong impact of climatic hazards (annual rainfall varying between 700 and 1250 mm, variable arrival date of the first rains) on agricultural practices. The many interviews we have conducted with farmers in this region confirm this: the climate's random nature is a much greater obstacle to development—and one that is more keenly felt—than that caused by long-term climatic upheavals. This randomness, which has already been observed for a long time, is likely to increase in the future, as is the intensity of violent events, which has also already been observed.

2.1.4.2 In Iringa on the highlands of southern Tanzania

Data available from the Iringa meteorological station allow for a comparison of annual rainfall volume over the 1972–2014 period (Fig. 2.5). Not much marked evolution can be deduced from this, except for a decrease in the 2000s, whose trend seems to be reversed at the beginning of the following decade.

The rains' irregularity is, again, a significant fact. However, in interviews with older farmers, the shortening of the rainy season seems to have been the most significant change. "In the past,"

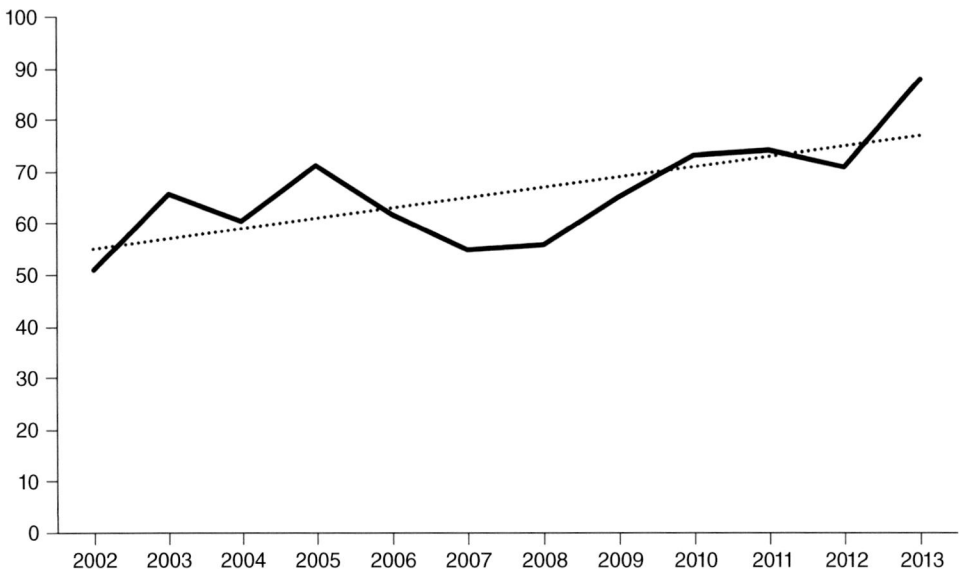

Fig. 2.4. Evolution of rainfall volume on the rainiest day of the year in Mpika (2002–2013). (From Thomazo, 2014, based on data collected at the Mpika weather station.)

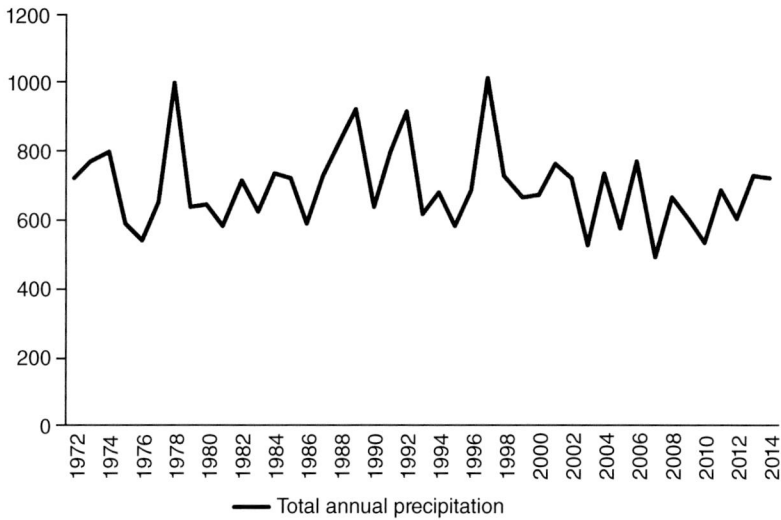

Fig. 2.5. Evolution of rainfall volume in Iringa (1972–2014). (From Verhoog, 2015, Iringa weather station.)

they say, the rains started a little earlier than they do now and ended in June, so the rainy season was almost 2 months longer (May and June) than it is now. Corn could be planted from January to March and harvested from August to September, a much longer growing season. It should be noted that this change would have taken place as early as the early 1970s (although the period that followed became more favorable) and is not perceived by farmers as a recent phenomenon. We will see that the earlier interruption of the rains considerably strengthens the interest in areas where it is possible to have two cycles per year: lowlands or small, irrigated perimeters.

The fact that the period of climatic deterioration (shortening of the rainy season) in the early 1970s corresponds, in Tanzania, to the villagization policy period, which brought major disruptions, complicates this analysis. In the memory of those who lived through this period, the deleterious effects of climatic deterioration on one hand and forced habitat displacement on the other combine to increase farmers' difficulties. In this case, this is one of the major difficulties that arise when we talk about climate change and it is very difficult to distinguish between the effects of climate change on one hand and the effects of public policies on the other.

2.2 Change and Vulnerability: Lessons from History

2.2.1 The historical confinement of Zambian agriculture to corn monoculture

Corn accounts for about half of Zambia's food supply, which, along with Malawi, are the two countries where corn occupies the largest share of the population's diet, ahead of Zimbabwe (about 40%) and South Africa (Douillet, 2013).

This crop's importance in Zambia has been the center of much of the attention paid to corn in the agricultural sector. Corn plays a central role in the country's agrarian landscapes, feeding the population and in state agricultural food policy. In terms of policy surrounding corn, Zambia has subsidized a technical package of improved seeds and fertilizers, built up strategic stocks, and controlled exports.

The dominant character of this crop and the preponderance of a cultivation method that favors sole cultivation and monoculture are the result of nearly a century of agricultural development almost entirely devoted to the exclusive promotion of this technical model. There is no doubt that the fragility of Zambian rainfed agriculture and its vulnerability to both climatic hazards and possible long-term climate change are partly the result of these choices.

The spread of corn cultivation over a large part of the African continent is a striking example of the long-standing and profound transformations in African agriculture. However, the diffusion of this innovation across the continent has taken different paths and has resulted in contrasting situations today.

The irruption of plants of American origin (corn and *Phaseolus* bean) into African agrarian systems began at the end of the 16th century when the first seeds from the New World were sown on the coasts of the Gulf of Guinea or Angola. Only two centuries later, they were known, adopted, and sometimes massively used in certain regions, such as in the hills of the Kingdoms of Burundi and Rwanda. In these countries, for example, these new crops' integration into farmers' work calendar contributed to the emergence of a new agrarian system that was twice as productive as the previous one, at the end of an unprecedented agricultural revolution (Cochet, 2001), an agricultural revolution that is in every way different from the Green Revolution that some countries of the South experienced in the second half of the 20th century. Indeed, this new plant material has enriched the farmers' array of resources and has made it possible to:

- increase agrobiodiversity
- generalize the practice of intercropping
- multiply crop cycles on the same field
- encourage the progressive filling of farmers' work calendars
- decisively increase labor productivity (the amount of food produced per laborer is doubled)
- significantly reduce the risks involved by allowing the multiplication and staggering of harvests (Cochet, 2001)

However, the predominant place that corn occupies today in Southern Africa in general and in Zambia in particular is not due to this long historical process of adopting new plant material from the New World. It is mainly due to the agricultural policies that, in these countries, have made this crop the pillar of agricultural development. This massive public support did not begin with the independent government of Kenneth Kaunda, as is sometimes claimed, but was already a central element of the policy of the preceding periods' White governments. Corn was already present in "indigenous" agriculture as a mixed crop, in complex cropping associations that usually combined cereals, legumes, and root and tuber crops.[5] However, sole corn cultivation was intensively promoted by the authorities of Southern Rhodesia and later by those of the short-lived Federation of Rhodesia and Nyasaland (1954–1963)[6] for the exclusive benefit of White settlers. While genetic selection efforts, conducted in particular at the Salisbury research station, worked to satisfy the needs of European farmers, "African" agriculture was the responsibility of the Department of Native Affairs (McCann, 2005). Its mission was not to understand farmers' practices but to change them as quickly as possible. This is why genetic research is carried out without taking into account the corn populations that had been adopted, and then selected, by several generations of Black farmers according to their needs and the ecosystems to which they had access. The hybrid (white corn) varieties developed are rather adapted to the industrial production of corn flour for the preparation of the dough that will become Black workers' staple food, especially in the mines. This production was highly subsidized and protected from the 1930s onward by the establishment of (i) collection and storage infrastructures exclusively for White farmers' production, (ii) differential prices for White farmers' benefit, and (iii) restrictions on the marketing and circulation of corn produced by Africans (McCann, 2005, p. 147).

The resumption of genetic improvement work after the Second World War and an effective support policy led to a spectacular increase in white corn production in this region of the world from the 1950s onward, even before the Green Revolution started in Asia (idem). The hybrid varieties selected at that time,[7] in order to achieve their yield potential, needed to be sown as early as possible. This was only possible by completing tillage even before the beginning of the rainy season (which required a tractor), and by falling back on the use of synthetic fertilizers and the annual purchase of hybrid seeds: all elements that were little or not at all accessible to Black farmers (idem).

Independence and Kenneth Kaunda's arrival in power marked a major break. Without hindering the development of large White-owned farms, corn production was promoted among small-scale producers. In order to overcome the obstacles to the mass adoption of the technical package associated with the cultivation of hybrid white corn varieties, an input subsidy program was put in place with guaranteed purchase prices for producers. In addition, local genetic research produced new hybrid varieties better adapted to family production systems: shorter-cycle varieties that could be sowed later, and double or triple hybrids that could be reseeded for one or two additional years without significant yield losses (McCann, 2005, p. 165). National corn production thus increased rapidly, especially in the 1980s (Fig. 2.6).

Was it a social breakthrough with widespread benefits? No doubt, but what the technical model promoted, despite a better adaptation of the varieties offered to small producers, was in line with previous decades: promotion of sole cultivation (and row seeding) and continuous corn/corn succession on the same fields (monoculture) with subsidized synthetic fertilizers. In addition, the extension system put in place relied on model farmers chosen according to their dominant social position and, it is believed, likely to facilitate the innovation's dissemination; a pyramidal model of extension in every way comparable to those put in place across the continent and still reproduced today (notably, alas, in the context of the extension of so-called "conservation agriculture").

In the Mpika region, for example, which was then dominated by slash-and-burn agriculture (Miombo type) and the cultivation of finger millet, cassava, peanuts, and beans, L. Thomazo explains that the "model" families had their land cleared by tractor and then, with the help of day laborers, installed a continuous crop of sole corn with synthetic fertilizer. From 1985, the support program was extended to all farmers, but with a limit of two limas per producer (0.5 ha).

The adoption of technical recommendations for corn cultivation became the key to accessing fertilizer at affordable prices. From then on, corn acreage, yields, and production fluctuated according to the government's ability to provide the necessary fertilizer and seed in a timely manner and to purchase the production offered at the advertised price. As is well known, the end of the 1980s also marked the abrupt end of subsidies (crisis and structural adjustment). It led to the immediate collapse of production (Fig. 2.6), with an immediate and devastating impact on all farmers whose cash income was still essentially based on the sale of corn. Families were faced with two challenges: replacing corn in their diet, at least partially, and finding an alternative source of cash income (Thomazo, 2014).

Given the negative consequences of the end of subsidies, a public support system was restored in the early 2000s. It was once again based on a program of input subsidies (Farmer Input Support Program, FISP) and guarantees concerning the purchase price to producers and the sale of their production (Food Reserve

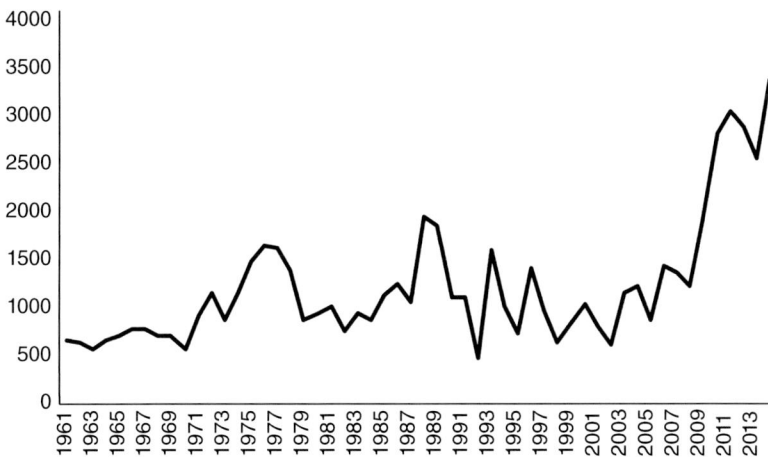

Fig. 2.6. The impact of the input subsidy policy on annual corn production in Zambia in thousands of tons. (Courtesy of FAOStat.)

Agency, FRA). Despite criticism of the policy, particularly with regard to its weight in the Ministry of Agriculture budget, it has been successful, as in neighboring Malawi, in increasing production and food security at the national level (Fig. 2.6).

The restoration of the corn subsidy system, however effective it may have been in increasing production, did not call into question past choices, in particular the preference for sole, and often continuous, corn cultivation. Was corn the crop best suited to the fairly dry tropical climate of the Zambian highlands? Was corn the crop best suited to the sandy soils—with low soil water reserves—that cover much of the country?

Although farmers have locally adapted the technical message to their particular conditions and have sometimes moved away from the dogma of sole cultivation, the fact remains that the preponderance of corn in the production system makes them undoubtedly vulnerable to the many hazards that weigh on production. There are climatic hazards, of course, but above all are hazards linked to the farmers' dependence on the arrival of fertilizers and the ability of the public authorities to subsidize their purchase, and there are also hazards linked to the possible sale price of corn, which is itself linked to the purchasing capacity of the Food Reserve Agency (FRA). It can, therefore, be seen that while climatic hazards concern all cropping systems and all production systems, the risk incurred is much greater for corn grown as a sole crop, as it has been popularized since independence, and for production systems centered around this crop.

2.2.2 Tanzania: villagization and corn technical package

The policy of settlement regrouping implemented from 1974 onward under the presidency of Julius Nyerere had serious consequences in this region, as in many Tanzanian regions, especially since the policy was implemented in the middle of a period of climatic deterioration (an unfortunate combination of circumstances). Forced villagization is said to have caused a sudden drop in agricultural production (estimated at 30%), which the political authorities attributed to drought (Raison, 1982). Just as today, many difficulties in the agricultural sector are attributed to climate change in order to avoid questioning the agricultural models promoted. For those interested in the interrelated consequences of climate change and public policies on agriculture, this period is rich in lessons, provided that, in order to understand it, we look back on the period that immediately preceded it.

2.2.2.1 Dispersed settlement, access to diverse ecosystems, and intercropping: an overview of the agrarian system before villagization

In the Kiponzelo region, not far from Iringa, historical interviews conducted by Niel Verhoog have allowed us to reconstruct with precision the type of farms that were most common at the time. Each production unit occupied a "strip" starting at the top of the hills and ending in the wet depressions below (Fig. 2.7).

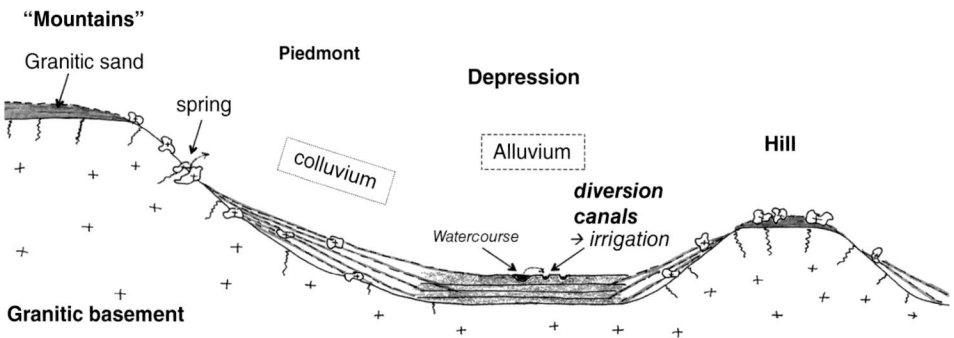

Fig. 2.7. Typical toposequence of the Kiponzelo region in the southern highlands of Tanzania. (Illustration by N. Verhoog.)

The dispersed habitat was located on the lower hillsides, often below the springs emerging near the escarpment's break-in slope. Fields of rainfed crops surrounded the family home, sometimes delimited by lines of bamboo. Corn, different varieties of beans, and a few "vegetables" (amaranth, radishes, tomatoes) were cultivated in association, in an increasingly dense manner as one moved closer to the house, a source of fertility (household waste, ashes from the fireplace). In addition to domestic waste, the renewal of fertility was ensured by the few years of fallow land (shrubbery) that alternated cultivation years in the rainfed fields, but also by the spreading of manure from livestock (cattle, goats, pigs, poultry), widely practiced at the time.

In the forest massifs (escarpments, sparsely inhabited interfluves), finger millet was cultivated over small slash-and-burn areas, then transformed into flour or *pombe*, the traditional beer. This slash-and-burn cropping system (known locally as *mahonyo*) was adapted to the low biomass level of the open forest (Miombo type) and the presence of an almost continuous herbaceous layer (grass), also characteristic of open forest.[8]

In the wet depressions that dotted the lowlands, "lowland" fields were cultivated in the dry season, while others could be irrigated from springs on the hillsides. It should also be remembered that the rainy season at that time was much longer than it is today, with rains lasting until late May or even early June.

The size of these production units ranged from 3 to 4 acres for the smallest to 10–25 acres for the largest, with herd sizes also reflecting the farms' socioeconomic differentiation.[9]

2.2.2.2 Villagization and its consequences

Implemented between 1974 and 1976 in the region, villagization caused a profound destabilization of the agrarian system. It resulted in a significant increase in farmers' vulnerability, particularly with regard to climate change that occurred at the same time. From one day to the next, farmers were ordered to move to the new "centers" and build their main permanent dwellings there. The houses of the recalcitrant were burned down, and the occupants were forcibly moved to the new spaces made available to them (between 0.25 and 0.5 acres). The consequences of this untimely displacement of housing were as follows:

- It was now necessary to walk or cycle to the fields every day, which increased daily transport time, reduced working time available on site, and thus reduced the exploitable agricultural area per laborer.
- The movement and herding of livestock became more complicated due to the obligatory concentration of livestock in the new villages, which also increased the risk of transmission of parasites and diseases. Epizootics, lack of care, and increased difficulties in herding led to excess livestock mortality.
- Renewal of fertility was made difficult by the death or sale of livestock. In addition, it was now necessary to have a cart to transport dry cow dung from the corral to the field and therefore to have oxen, which became increasingly difficult.
- The cultivated areas and the yields decreased by almost 20%.

Thus, while the climate change under way in those years imposed a tightening of farmers' work calendars and widening of the bridging time, the policy of compulsory grouping of housing placed farmers in great vulnerability, preventing them from coping with climate change in conditions that allowed adaptation. In addition, the obligation to cultivate certain fields collectively within the framework of the newly created cooperatives also contributed to diverting part of the labor force to fields that were ultimately not very productive.

However, other agricultural policy measures, decided at the same time as villagization, could have improved farmers' lot by helping them cope with climate change: construction, in each village, of antiparasitic baths for cattle; credit distribution of seeds and fertilizers to farmers (via cooperatives); guaranteed disposal of corn production (also via cooperatives); and dismantling (after independence) of colonial tobacco plantations with reallocation of land to farmers.

But these efforts have themselves been wiped out by the deleterious effects of villagization.

2.2.2.3 From the FAO technical package to the One Acre Fund program

After the "National Corn Campaign" based on Green Revolution principles and launched by the World Bank in 1975, it was under the aegis of the FAO that a new corn technical package was disseminated in the region starting in 1985. The FAO proposed several types of "hybrid" corn in a "technical package" that farmers could purchase, as a group, on credit. The package included five bags of fertilizer and 4 l of seed per acre. The success of this technical package was very limited: its promoters did not mention the crops that had to be abandoned (associated with corn in the "traditional" case), nor the early harvesting of fresh corn at the time of the bridging time (now prohibited), nor the cost of fertilizers and seeds.

Since the establishment of the FAO promotion program in the 1980s, the promotion of corn cultivation has been at the heart of Tanzanian agricultural policies, particularly in the Iringa region. Today, even as farmers seek to diversify their production systems by developing more fields in the lowlands (residual moisture) or new irrigated perimeters, the bulk of the extension apparatus—even if it is now largely subcontracted *de facto* to NGOs and private investors—is still focused on improving corn yields.

This is the case, for example, of the *One Acre Fund* (OAF) program proposed in 2014 by an American organization. The technical package includes, in rainfed cultivation and for 1 acre: coated hybrid seeds, 50 kg of DAP (18-46-0) applied in the planting hole using a measuring spoon provided by the program, 50 kg of KEN fertilizer (whose composition should be specified) applied at two points (three-leaf stage and flowering), and finally, 10 kg of urea.

Farmers emphasize this program's advantage in terms of access to credit, seeds, and quality synthetic inputs. In addition, the fact that the loan can be repaid in successive stages and with no minimum amount is appreciated by farmers.

However, although the yield is good, provided that climatic conditions are favorable (seven to eight 100 kg bags of grain/acre), the cost of this technical package appears prohibitive. In 2015, seeds and fertilizer, delivered on credit, cost 210,000 Tanzanian shillings (TZS),

but in addition, urea, pesticides (insecticides), and transportation of the crop from the field to the house had to be purchased. The total cost was estimated at TZS 260,000 (plus tractor service and mechanical ginning) with an estimated production value of TZS 320,000 in the best case (i.e. a high price of TZS 40,000 per bag).[10]

Thus, in the event of a bad harvest, the farmer will have to find the money needed to repay the loan elsewhere (at the risk of having his assets seized as collateral). We do not have enough time to evaluate the impact of this program (implemented in 2014 in the region studied), but similar experiments conducted in Ethiopia over the past 20 years (the Sasakawa/ Global 2000) call for caution: they have been proven to be dramatic for many farmers who were unable to repay their loans after an unfavorable year in terms of climate (Planel, 2008; Cochet, 2009).

2.3 Diversity Reduces Family Farms' Vulnerability to the Hazards of Rainfed Agriculture

Despite the impact of policies promoting "intensive" corn cultivation, corn is not always grown as a sole crop—far from it. Although the crops used in intercropping with corn—most often the "American trilogy" with *Phaseolus* beans and squash—are more discreet than corn itself and therefore not very visible in the landscape (one has to enter the field to discover them), they are nonetheless quite widespread, especially on small family farms.

2.3.1 In Northern Zambia (Mpika), lessons from the chitemene cropping system and its contemporary evolution

The role of corn in family farming in the northern highlands of Zambia appears to be less hegemonic than it might seem. A wide range of crops are grown, mostly for food, with surplus crops sold to provide the bulk of cash income. Here, the old slash-and-burn production system, adapted to the *miombo* (open forest with *Brachystegia* and *Julbernardia*) is still widely practiced. The principles of this cropping system—known as *chitemene*[11]—deserve to be recalled here because contemporary developments in this agriculture cannot be understood without clear references to "starting point."

2.3.1.1 The chitemene cropping system

This is a slash-and-burn cropping system adapted to the open forest's low biomass level and the presence of an almost continuous herbaceous layer (grass), also characteristic of the open forest.

As a result, fertility reproduction and the destruction of herbaceous cover (and the corresponding seed stock)—two main functions of tree regrowth and the basis of slash-and-burn cultivation's agronomic logic—cannot be ensured by tree regrowth itself, even if its duration is fairly long.

This is why farmers in this region of the world have developed an original practice that is typical of this environment and relatively efficient. Of the surface area cut down each year, only a small part is cultivated. After the felling, the cut branches are transported to the center of the felled field and piled up in a pyre on a space much smaller than the felled field which will be the only one effectively sown. The concentration of the biomass in a smaller area allows for two things: concentrating the fertility—and thus regaining the efficiency of a much denser tree regrowth—and intensifying the burning, the latter allowing for the partial destruction of the herbaceous stratum's grass clumps (destruction completed by a long weeding operation). Moreover, this (manual) displacement of biomass toward the center of the field also modifies the felling technique. As the trunks could not be moved easily, they were not felled: they climbed the trees to prune only the branches that could be transported. The result is a singular landscape and a very clear differentiation in the dynamics of the regrowth that will later take place between the center of the field, which is effectively cultivated, and the periphery, which is pruned but neither burned nor cultivated (see Photo 2.1):

- In the center, where fertility is concentrated, regrowth will be delayed by several years of cultivation (at least three) and the annual cutting of stump sprouts. The large trees, pruned at the time of felling, are killed

by burning, which is particularly intense because of the accumulation of biomass. Their dry trunks signal from afar the cultivated (or formerly cultivated) heart of a *chitemene* field.

- On the periphery, regrowth is immediate: there is no burning or any other hindrance to the emergence and immediate development of shoots, the herbaceous layer taking advantage of this temporary increase in light to become denser. This ring is also characterized by the regrowth of large trees (only pruned) and their characteristic silhouette. On closer inspection, it can be seen that vast areas of *miombo* are today composed only of trees that have undergone at least one phase of pruning related to *chitemene*, which testifies to the widespread and sustainable (under certain conditions) nature of this practice.

L. Thomazo (2014) has carefully reconstructed the intercropping and crop successions characteristic of this mode of environmental exploitation: in the first year of cultivation, and without prior tillage, the crop is always finger millet, associated with cassava, sometimes also squash and a few subplots of corn; this association is followed by a crop of peanut or *voandzou*, itself associated with the cassava cut the previous year and still in place. The cassava remains in place alone the third year, during which it will be gradually harvested; a crop of bean or peanut may then come to extend this cropping sequence for one, two, or three more years.

2.3.1.2 When there is a lack of space

The system known as *chisebe* ("poor man's slash-and-burn") is an evolution of slash-and-burn cultivation toward shorter-duration fallow land systems (mainly herbaceous and with a low biomass level), similar to a savannah, and which requires the use of tillage. It is conducted in the following manner: due to the lack of sufficient available area, the principle of concentrating branches on the part that will be cultivated is abandoned. The product from felling is gathered in small piles which, after burning, will receive certain crops (potatoes, squash, cucumbers, watermelons), while the rest of the field is horizontally plowed with a hoe. Cassava cuttings are first planted on the whole field; finger millet is broadcast sown in the field afterward (also on the whole field). In the second year, cassava and peanuts are associated with corn being grown only from the third year onward and with other plants (beans or peanuts), continuously for 3 or 4 years. This can lead to a shift to a ridge-cropping system, where tillage and weeding are heavy tasks.

2.3.1.3   Corn technical package and climatic hazards

The policies promoting corn cultivation (*supra*) encourage farmers to start growing corn continuously, which, in addition to the availability of fertilizer, requires heavy tillage: a real break with the *chitemene* system. The *chisebe* cropping system (described above), after 2 or 3 years of associated cropping, then evolves toward corn monoculture. The history of the last 50 years seems to play out according to a shifting balance between *chitemene* (never completely abandoned), *chisebe*, and continuous corn cultivation.[12] This balance depends above all on the availability of land and access to fertilizer, the latter varying according to the period (*supra*) and the family. When fertilizer subsidies were discontinued (especially in the late 1980s), many returned to their *chitemene* fields, which served as a "safe" and therefore antirisk mode of farming.

In the area studied in this research program (Katongo Kapala, Mpika District), L. Thomazo studied in detail the impact, in terms of yields, of the various climatic hazards identified (late arrival of the first rains, drought episode immediately after the first rains, violent storms) on the different cropping systems. This analysis shows that the yields from intercropping systems—*chitemene*, *chisebe*, corn + bean association—are often less affected by climatic incidents than those of sole cropping systems. In addition, it was noted that the rainfed corn yield, regardless of the cropping method, is strongly influenced by the rains' arrival date or possible drought episodes. The farm's vulnerability therefore increases as (1) corn's place in the cropping pattern increases and (2) the share of associated crops decreases. This is why farms whose main production is corn in monoculture suffer greatly from the drop in corn yields in years when the first rains are late.

In the Mkushi region, much closer to Lusaza and studied in detail by Esther Laske (Laske, 2014), the "all-corn" approach also resists reality. With the exception of large, single-holding agricultural enterprises (farm blocks in place since the 1950s) under irrigation center pivots, the landscape has yet to be artificialized and is largely dominated by the more or less degraded *Brachystegia* open forest (*Miombo* type). Between the lowland areas, which are still not very cultivated, and the crest lines of the main interfluves, the hillsides with moderate slopes are the preferred places for cultivation. In the lower part, in the vicinity of the lowlands, there is a ring of forest where hamlets are located (the undergrowth is carefully cleared). Then, in the houses' immediate vicinity, there are small fields of associated corn (squash, beans, and voandzou), a few ridges of sweet potato, clumps of banana and mango trees, and so on. As we move up the hillside, the fields become larger and often consist of sole crops: corn, soybeans, peanuts, or more rarely, finger millet or cassava. On May 14, 2014, A. Axani met with us on one of these fields and explained how, after plowing with draft animal power, he combines corn, sown in rows, with squash and beans,[13] which does not prevent him from using the coated seed and fertilizer kit distributed under the *Farmer Input Support Program* (FISP).

Finally, whether in the Mpika region in northwestern Zambia or in the Mkushi region, farmers supplement their systems, when possible, with small, irrigated fields devoted to crops for sale and/or for self-consumption. Thus, the families with the highest incomes and the best ability to cope with the various hazards to which they are exposed are always those that can combine rainfed and irrigated agriculture in different areas.

2.3.2   The importance of having a diversified environment: the example of Kiponzelo (Iringa, Tanzania)

Although generally tabular on a small scale (the regional scale), the Iringa region's relief is locally quite accentuated: an escarpment approaching 2000 m in altitude in the west, intermediate hills, and grassy "depressions" around 1650–1700 m.

Numerous watercourses and "springs" from the heights have allowed the planning of small, irrigated perimeters. The ecosystems to which farmers can have access are thus very varied, in terms of both soil composition and the ways they can be used (Fig. 2.7).

The different landscape units are therefore as follows:

- Western escarpment: although the soil may be suitable for cultivation in places, the difficulty of access to this area hinders cultivation. Large areas of pasture dominate here and allow herders to send their herds to graze during the rainy season.
- Western escarpment convex slopes: below the hillside's upper part, the slope is steep and cluttered with granite blocks; a break in the slope marks the transition to the hillside's lower part. Numerous springs appear at the slope's break and feed small, irrigated areas. This was also the favored place for habitation before the villagization carried out in the 1970s.
- Lower hillsides: the slopes become progressively gentler (colluvium) as they approach the lowlands and are intensively cultivated, mainly with rainfed crops (associated corn). Many "useful" trees dot the landscape and provide shade for workers, as well as additional fertility (leaf fall, legumes). Between the fields, often to delimit the properties, is a mesh of bamboo mounds forming a hedgerow landscape. Bamboo sap is harvested to make a popular alcoholic drink. There are also small, gravity-irrigated perimeters (see Photo 2.2). The small canals built between fields in the direction of the slope are regularly blocked by small weirs. These weirs both slow the downstream water flow and raise the water table's level in the fields on either side (lateral infiltration). The fields are partly cultivated during the rainy season (notably sole corn), but mainly during the dry season for market gardening: Chinese cabbage, black radish, peas, beans, tomatoes, onions, peppers, eggplant, garlic, carrots, spinach, and so on.
- Wet lowland areas: these areas are cultivated in the offseason or with two crop cycles per year, planned in raised beds. These are clearly distinct from the small, irrigated perimeters located on the hillsides. However, they correspond to a similar type of planning in many respects (morphology, shape, and size of the fields, crops, and crop succession), and the same term is used to designate them: *vignungu*. Two types of lowland fields can be distinguished. Where the water table does not risk submerging the crops (close to the edge), fields are created to be cultivated all year round. Below, where the water table is higher, the fields in the form of mounds or "slabs," are more rounded and raised. They will only be cultivated in the dry season (see Photo 2.2). Complementarity with the piedmont fields, a slight shift in agricultural operations, and specific crops (corn on the cob sold fresh, for example) make these lowland fields highly prized.
- Year-round wet depressions: These are generally flat areas, but their humidity increases from the periphery to the center. These areas are used for dry-season grazing and are often communal.
- Interfluves: These are flat-topped hills and convex hillsides with a very gentle gradient, with sandy soil on granitic sand. These interfluves are themselves to be subdivided between the summits still covered with clear *Brachystegia* and *Julbernardia* forest (*miombo* type), which are strongly degraded, and the lower hillsides cultivated during the rainy season. Here also, small, irrigated perimeters have sometimes been planned by diverting water from a stream.

When reading about this layered and diversified landscape, one realizes how important it is for a farmer to have access to as many of these ecosystems as possible. Not only does this access multiply the realm of possibilities, but the combination of several crops and cropping systems whose demands on work time do not completely coincide (especially when rainfed and off-season crops can be grown together, of course) also makes it possible to increase overall labor productivity (the net value-added production per laborer) by saturating the family labor force, which is available throughout the year. Access to irrigation appears to be a key element in this process. In addition, irrigation also provides access to higher value-added crops due to their greater market value.

2.4 Access to Small-Scale Irrigation, Key to Increasing Diversity and Raising Labor Productivity and Farmers' Income

2.4.1 In Mpika district (North Zambia)

The Katongo Kapala region (Mpika district) has a fair number of irrigation canals (with simple water intakes on existing rivers) that were planned on the initiative of small farmer groups starting in the 1960s. However, not everyone has access to irrigation. An excerpt from the typology presented by L. Thomazo (2014) for this region gives the following results:

- The worst-off families are those who have only had access to small (<1 ha) fields of savannah (one accessible ecosystem) where they have attempted to specialize in corn production with government support. Relying on the assurance that the sale of corn will bring in some money and, having few other financial resources, they focus on establishing a small intercropping field: corn + bean or peanut + pumpkin + sweet potato, in rotation with sole legume crops. Although the value added per land unit is relatively high, thanks to intercropping, the income is very low (the equivalent of 80–100 euros/worker/year) due to the farm's small size and well below the poverty line.
- Families with wide access to open forest still practice *chitemene* (*supra*). Yields are good and largely provide the family with cassava, pumpkin, finger millet, peanuts, and beans. Some of the finger millet can even be stocked and sold later. In addition, these families practice shorter rotations in the less forested parts of their property (the *chisebe* system). In this way, they both ensure a minimum of cassava for the family's diet and open fields where a bean/corn/peanut/corn rotation can be established after the cassava harvest, benefiting from state corn subsidies. The farm's location in the heart of the open forest also facilitates small-scale livestock production (goats and poultry), where the necessary fodder resources can easily be found. These farms do not have access to irrigation, but the combination of different activities and the fact that the two cropping systems practiced do not mobilize the family's labor force at the same time allows these families to generate an income slightly above the poverty line, in the order of 200–250 euros/worker per year.

- As soon as producers have access to small irrigable areas and can set up a garden for corn, tomatoes, and other vegetables, in addition to the other cultivated fields, their situation improves significantly. This is the case, for example, for (i) small farms that generate income of about 300 euros per worker per year with only 0.25 ha irrigated to complement the other fields, or (ii) slightly larger farms that, with 2–3 *limas* (0.5–0.75 ha) of the irrigated garden, generate income of 500–600 euros/worker/year. These incomes can be increased if livestock farming is added to the production system.

In addition to the very positive impact of irrigation on farmers' incomes (albeit limited by difficulties in accessing the market), irrigation also makes it possible to better withstand possible drought episodes during the rainy season by providing supplemental irrigation.

2.4.2 In Mkushi district (Zambia)

This region has seen significant irrigation planning (diversion canals and gravity irrigation, then direct pumping in the Mkushi River) on large farms established in the 1950s for the benefit of White farmers (farm blocks), on the right bank of the Mkushi River. From the 1980s onward, irrigated market gardening developed rapidly on small family farms on the left bank. Tomato production, which is now the dominant cash crop in the region, has contributed significantly to the increase in farmers' income.[14] The means available to farmers to mobilize water resources and to irrigate their fields are derisory and constitute a severe limit to the planning of irrigation. An excerpt from the typology presented by E. Laske for this region gives the following results:

- 90% of families in the region only have watering cans. The irrigated garden is therefore very small (<0.1 ha) and installed in the lowlands where water can be drawn out by digging a simple hole at the edge of the field and using a watering can. The garden is mainly devoted to tomatoes (2 or 3 cycles/year). With only manual tilling tools, these farmers can only cultivate small areas of rainfed crops (1–1.5 ha): corn, of course, but also finger millet, sweet potato, peanut, and bean in short rotation. The annual income generated per worker is in the order of 300–600 euros. This income can be increased if the farmer has access to draft animal power (owned or through a service provider) and can thus increase the dry-cultivated area (the irrigated area is limited by irrigation equipment).

- The families that have been able to build an elevator dam (+feeder canal) on a small secondary river have more water and are able to irrigate a larger surface area, in the order of 0.6–0.7 ha. They can then often grow three cycles of tomatoes/year, as well as other diversified market garden crops. The agricultural income reaches much higher levels, more than 1000 euros/family worker/year. Here again, it is the diversity of crops, both irrigated and rainfed, that guarantees the increase and stability of income.
- A very small number of families have a motor pump and can thus take the necessary water directly from the river. These are owner-managed farms that call upon the services of numerous day laborers throughout the year. The irrigated area reaches 1–2 ha (the total cultivated area can reach 10 ha, sometimes even more). Labor productivity is not higher than in the above-mentioned farms (except for watering), but the family's income can reach much higher levels insofar as the many pieceworkers (or permanent workers) are paid at a level much lower than their productivity.

Finally, this region is also characterized by very large salaried farms, moto mechanized and now specialized in annual crops (wheat/soybeans, for example) under center-pivot irrigation. They occupy the entire right bank of the Mkushi River and monopolize the vast majority of the available water resources. Although they generate a very high level of income for their owners, they create little value added per unit area compared to the small farms on the left bank, which are much more labor intensive.

2.4.3 The example of Kiponzelo, Tanzania

Here again, the limiting factor for vegetable production is access to areas suitable for off-season cultivation (lowlands, mountain springs, or rivers). These fields are strategic in that their access determines whether or not it is possible to produce vegetables, and thus to have crops with high value added. With access to the market and the possession of livestock that provides organic fertility, small-scale irrigation appears essential here.

Among the production systems identified and analyzed in detail by N. Verhoog, we will retain, for illustrative purposes, the following categories.

We note that the first category presents modest results due to limited access to cultivable dry-season fields. These farms combine three types of fields:

- First, the field around the house, exclusively devoted to food crops: corn, beans (dwarf and climbing), squash and sunflowers, banana trees, cassava, sweet potatoes, and amaranth. The little organic fertility available (household kitchen waste, poultry manure) will be dedicated to it, as well as, sometimes, a little synthetic fertilizer.
- Next, another field of rainfed corn, further from the house, where intercropping is also appropriate (beans and squash), at least on part of the field.
- Then come the lowland fields. During the dry season, these fields are cultivated with some beans, peas, radishes, and Chinese cabbage (watered by bucket). They are then sown with corn and beans before the rain starts again and thus watered again a few weeks before the rains take over.

On these farms, only a few bags of corn will be kept for self-consumption, with the rest of it being sold directly after the harvest to supplement a meager cash flow. In bad years, the farmer will have to buy corn to feed his family during the bridging time (February–March), a situation that can become critical if prices are high. The farms practicing this system are therefore vulnerable to climate change. They have a very low income—around 120 euros/worker/year, including self-consumption. Having access to very little cultivable land in the offseason, most of these producers work outside the farm (making bricks, day laborer on another farm, small-scale trading, etc.).

The second set of farms is slightly larger (3.5–4.5 acres). In addition to the field near the house, which is always present and where associated food crops prevail, these farms have larger dry-season fields. Fertilizer input is more important, which can be supplemented by organic fertility from a small livestock unit (one fattening pig, one sow, and her piglets). The Chinese cabbage crop is then largely intended for sale. It is sown in nurseries, then transplanted and fertilized with manure from the livestock pen. As for the corn crop, when "special care" is given to it (certified seeds, more fertilization and treatments, careful weeding), it is possible to harvest the largest fresh cobs in March–April–May and sell them directly to retailers. The agricultural income remains more than modest, around 200 euros/worker/year; it can be supplemented by an external activity (e.g. making bricks).

A third set of families have four to six acres, contract out draft animal power, grow potatoes and corn (sold fresh), and have slightly more livestock, especially pigs. Plowing with draft animal power saves time, but delays in sowing are frequent because farmers are still dependent on those who have draft animal power. Today, these farms have no cattle, whereas their parents often had cattle before villagization. Both "local" (farm variety) and "hybrid" (certified seed) corn are grown. The former is planted near the house, mainly for self-consumption, the latter for sale. A field of finger millet is still installed from August to May, the finger millet being intended in particular to be processed in the manufacture of *pombe*, artisanal corn beer. The economic results are better, and the family agricultural income is established at around 300 euros/worker/year.

A fourth set of families have slightly larger areas (5–7.5 acres) and are clearly distinguished from those described above by their relative specialization in dry-season-irrigated market gardening and the use of hired labor (day laborers). With a substantial cash flow to purchase the necessary seeds and inputs, these farmers are primarily concerned with pest and disease management. By enabling staggered sowing, transplanting, and harvesting, irrigation allows for a substantial improvement in farmers' income. The majority of the total gross value added, 60%, comes from market gardening. The economic results are higher and the income per family worker exceeds 500 euros/year.

Other farms also practice irrigated market gardening, but on a larger scale (10–15 acres) and with extensive hiring of day laborers. This type of farm has large areas with market gardening potential and several water sources (shallows, springs, rivers) that make it possible to maximize each crop's yield by allocating it to the ecosystem best suited to it. When there is a lack of water (fields far from the water source, reduced flow, or difficulty in watering the crops), mechanical treadle pumps (more rarely motor pumps) are used to compensate for it and secure crop yields (see Photo 2.3). The economic results of these farms are much better. The agricultural income per family worker reaches the equivalent of 1000 euros/year, due to the production of vegetables with high value addition (tomatoes, onions, peppers, etc.) and a large quantity of corn cobs sold fresh.

A sixth case consists of even larger farms. While practicing off-season market gardening, they are engaged in large-scale rainfed corn production, sold directly in the city (Iringa), and thus at a better price, which implies more complex organization (transport, commercial contacts, knowledge and monitoring of the market, etc.) and the sale of large volumes (partly purchased from neighboring farmers). Improved seed and fertilizer are used. Since the areas devoted exclusively to corn are very large, plowing with a plow is essential (oxen or hiring out tractors/operators). Access to the external market and the possibility of selling corn at a better price allow for higher economic results, but the latter remain modest: approximately 320 euros/worker/year. It should be noted that there are also larger farms in the region that are dedicated to rainfed corn, sometimes with a tractor. Their income is of course higher, although the value addition generated per unit of land remains low and highly dependent on the rain.

Finally, other farms have managed to keep or be allocated a herd of cattle. Thanks to the organic fertility they are able to obtain, the yields achieved and their cultivation capacities are clearly above those of the poorest farms.

2.4.3.1 Social management of water: the achievements of peasant organization

In the three regions studied here, in Zambia and in the highlands of southern Tanzania, farmers have been able to collectively organize themselves to plan and manage small, irrigated perimeters that are quite effective in terms of increasing and securing income. This is the case, for example, of the small Isupilo perimeter (see Photo 2.4), which is about 15 years old and groups together about 20 irrigators. Irrigation is done by gravity according to the same system observed elsewhere: the water circulates downward in the direction of the slope between beds, with small weirs to slow down the circulation and encourage infiltration. Some of them have distribution pipes to facilitate the change of water from gully to gully (see Photo 2.5).

Real social water management has been instituted: annual cleaning of the canals in June, and fines for absentees (5000 TZS); individual water intakes for each person downstream of the canal; unlimited water unless the irrigators located downstream lack it (no conflict, for the time being . . .); those who have fields above the canal that cannot be irrigated by gravity can nevertheless draw without limitation from the canal, but with a watering can.

2.5 Conclusion

The economic results obtained by farmers in the regions presented in this chapter are very low—a few hundred euros per worker per year—in line with those of most farmers on the continent who are engaged in predominantly rainfed and essentially manual agriculture. Although proactive policies to promote corn cultivation and efforts to improve varieties and disseminate fertilizers have sometimes been successful, when funding was available to effectively subsidize these inputs and when they were available in a timely manner, it is clear that it is not "all corn" that has led to a sustainable improvement in rural people's conditions. Those who are able to generate better and more regular incomes and who can protect their families from the various hazards faced are almost always those who implement diversified production systems that combine a wide range of accessible ecosystems with intercropping and small-scale irrigation. This diversity can not only be exploited by those with a minimum of equipment—hand tools in good condition for tilling, weeding, and harvesting, of course—but also small-scale watering equipment, draft animal power, carts, and so on. In addition, there is no doubt that easier access to improved varieties, synthetic fertilizers, and phytosanitary products can contribute significantly to improving farmers' situations and securing their income, provided that the use of these inputs does not lead farmers to over-simplify their cropping combinations and, a fortiori, to monoculture, which carries considerable risks and is particularly sensitive to the increase in climatic, commercial, and political hazards.

Notes

[1] They were identified thanks to the expertise and contacts of Clémentine Rémy, a consultant working for SOFRECO in the framework of the IDSP (Irrigation Development Support Project).

[2] White corn consumption accounts for 28% of the energy intake of the ration in Eastern and Southern Africa. This proportion reaches 50% in Zambia and Malawi. It is 25% in Tanzania (on the rise in this country) (Douillet, 2013).

[3] The latest IPCC report itself insists, after explaining the methods and models used, on the caution to be exercised with regard to these projections.

[4] The regression coefficient, which is very close to 0 ($r^2 = 0.08$), puts this result into perspective by clearly showing the superiority of the hazard's effect over these 20 years over any upward or downward trend.

[5] In the past, food crops were mainly based on intercropping, initially centered on sorghum (associated with beans of the genus Vigna in particular, voandzou, pigeon pea), then also on corn since the arrival in Central and Southern Africa of plant material of American origin (corn, *Phaseolus* bean, squash).

[6] The Federation of Rhodesia and Nyasaland included Southern Rhodesia (later Zimbabwe), Northern Rhodesia (later Zambia), and Nyasaland (later Malawi).

[7] In particular, the SR 52 hybrid.

[8] This cropping system is better known in Southern Africa as chitemene. It will be described in detail in the following pages.

[9] Until independence, the region also included a few large colonial tobacco plantations of several hundred hectares, located on the most fertile flat areas and near rivers.

[10] Survey done with Fausta on June 17, 2015 (N. Verhoog and H. Cochet).

[11] It has been described for Northern Zambia by several authors, including Audrey Richards from the colonial era (1939), Chidumayo more recently (1987).

[12] Continuous cropping is sometimes replaced by farmers with a corn/legume rotation (e.g. bean/corn/peanut/corn).

[13] Every 10 ft, he sows a corn seed, a squash seed, and a bean seed in the same hole.

[14] This cash crop's omnipresence among small-scale producers, intensive pesticide use, the very high variability of input and tomato prices, and dependence on the market are all issues today. For farmers, they amount to as much as hazards that make tomato cultivation a risky activity.

References

Ahmed, S.A, Noah, S., Diffenbaugh, N.S., Thomas, W., Hertel, T.W. *et al.* (2011) Climate volatility and poverty vulnerability in Tanzania. *Global Environmental Change* 21, 46–55.

Chidumayo, E.N. (1987) A shifting cultivation land use system under population pressure in Zambia. *Agroforestry Systems* 5, 15–25.

Cochet, H. (2001) *Crises et révolutions agricoles au Burundi.* INA P-G/Karthala, Paris, France, 468 pp.

Cochet, H. (2009) *L'agriculture éthiopienne face à l'accroissement du risque.* Présenté aux Dialogues Franco-éthiopiens «*Ethiopie: une société vulnérable au défi du risque climatique et environnemental*». Center Français d'Etudes Ethiopiennes, Addis Ababa, Ethiopia.

Douillet, M. (2013) Maïs en Afrique de l'Est et Australe: la sécurité alimentaire régionale liée à l'amélioration du fonctionnement des marchés. *Le Déméter*, pp. 205–226.

IPCC (2014) *Climate Change 2014: Impacts, Adaptation, and Vulnerability. Part B: Regional Aspects. Contribution of Working Group II to the Fifth Assessment Report of the Intergovernmental Panel on Climate Change.* Barros, V.R., Field, C.B., Dokken, D.J., Mastrandrea, M.D., Mach, K.J. *et al.* (eds). Chapter 22: Africa (Niang, I., Ruppel, O.C., Abdrabo, M.A., Essel, A., Lennard, C., Padgham, J. and Urquhart. P.). Cambridge University Press, Cambridge, UK, pp. 1199–1265.

Kanyanga, J., Thomas, T.S., Hachigonta, S. and Sibanda, L.M. (2013) Zambia. In: Waithaka, M., Nelson, G.C., Thomas, T.S. and Kyotalimye, M. (eds) *East African Agriculture and Climate Change: A Comprehensive Analysis.* IFPRI, Washington, DC, pp. 255–287.

Kilembe, C., Thomas, T.S., Waithaka, M., Kyotalimye, M. and Tumbo, S. (2013) Tanzania. In: Waithaka, M., Nelson, G.C., Thomas, T.S. and Kyotalimye, M. (eds) *East African Agriculture and Climate Change: A Comprehensive Analysis.* IFPRI, Washington, DC, pp. 313–345.

Laske, E. (2014) Dualisme agricole le long de la rivière de Mkushi, Province centrale, Zambie. Mémoire de fin d'étude, UFR Agriculture Comparée et Développement Agricole, AgroParisTech/AFD/IDSP, 71 pp.

Leichenko, R. and Silva, J.A. (2014) Climate change and poverty: vulnerability, impacts, and alleviation strategies. *WIREs Climate Change* 5, 539–556.

McCann, J.C. (2005) *Maize and Grace. Africa's Encounter with a New World Crop 1500–2000.* Harvard University Press, Cambridge, Massachusetts.

Mulenga Bwalya, S. (2010) *Climate Change in Zambia: Opportunities for Adaptation and Mitigation through Africa Bio-Carbon Initiative.* Center for International Forest Research, Southern Africa Regional Office, Lusaka, Zambia.

Planel, S. (2008) *La chute d'un éden éthiopien: Le Wolaita, une campagne en recomposition.* IRD (à travers Champs), Paris, France, 430 pp.

Raison, J.P. (1982) *Les erreurs géographiques de l'ujamaa tanzanienne.* ORSTOM (Tropiques, lieux et liens), Paris, France, pp. 402–420.

Richards, A.I. (1939) *Land, labour and Diet in Northern Rhodesia. An Economic study of the Bemba Tribe.* International African Institute, Oxford University Press, London, 425 pp.

Thomazo, L. (2014) Diagnostic Agraire à Katongo Kapala et dans ses environs, Mpika District, Zambie. Mémoire de fin d'étude, UFR Agriculture Comparée et Développement Agricole, AgroParisTech/AFD/IDSP, 119 pp.

Verhoog, N. (2015) Diagnostic agro-économique, un système agraire autour de Kiponzelo (Iringa, Tanzanie): Adaptation au changement climatique. Mémoire de fin d'étude, UFR Agriculture Comparée et Développement Agricole, AgroParisTech/AFD, 119 pp.

II

Rice Cultivation in Flood-Prone Areas: Facing the Hazard

3 Managing the Combined Hazards of Rainy Season and Flooding in the Absence of Planning: Agriculture in the Flood-Prone Areas of Kilombero and Lower Rufiji (Southern Tanzania)

Hubert Cochet[1]*, Jean Luc Paul[2], Céline Tewa[3], and Philippe Le Clerc[4]
[1]*Comparative Agriculture Training & Research Unit/UMR Prodig, AgroParisTech, Palaiseau, France;* [2]*Associate Professor of Anthropology, Institute of African Worlds, University of the French Antilles, France;* [3]*Founder of the Zero Waste 'Les Gamelles' network;* [4]*Rural development and Food security Program Manager, European Commission, Delegation of the European Union to Mauritania*

Two regions were identified to address the case of flood-prone, rice-growing areas with low levels of planning: first, the lower Rufiji River Valley in southern Tanzania, and second, the Ifakara region on the Kilombero River, a tributary of the Rufiji (Fig. 3.1). These two small regions offer an excellent opportunity to reflect on how farmers can consider the risks associated with climatic hazards. Farmers are confronted with two types of hazards, both related to climate: first, rainfall volume and distribution (date of the first significant rainfall, distribution of subsequent rains), and second, flood volume and timing (location, duration), on which the areas harvested, effectiveness of fertilization (deposition by the flood), and yields obtained depend. These regions are also part of the development "corridor" advocated by SAGCOT (Southern Agricultural Growth Corridor of Tanzania), where different agricultural models (family farming and large-scale investment projects, including rice cultivation plans) are combined.

3.1 Rufiji River Flood Valley and Tributaries in Southern Tanzania

Projections of possible rainfall change in East Africa are relatively uncertain and do not make for a very clear trend (Chapter 2, this volume). Although southern Tanzania is one of the regions for which a rainfall increase is considered likely (Kilembe *et al.*, 2012), we will see that recent rainfall trends in the two regions studied call for caution: a clear decrease in rainfall is well documented in the Rufiji River's lower valley, but upstream, in the valley of its tributary, the Kilombero, the trend seems to be more upward (see below). This region of East Africa is also one that would be affected—all the models converge on this point—by an increase in the frequency and intensity of extreme events, particularly floods. The years 1997, 2007, and 2014 bear witness to these extreme events.

*Corresponding author: hubert.cochet@agroparistech.fr

©2024 CAB International. *Agrarian Systems and Climate Change: Journeys of adaptation in the Global South* (eds H. Cochet, O. Ducourtieux, and N. Garambois)
DOI: 10.1079/9781800628137.0003

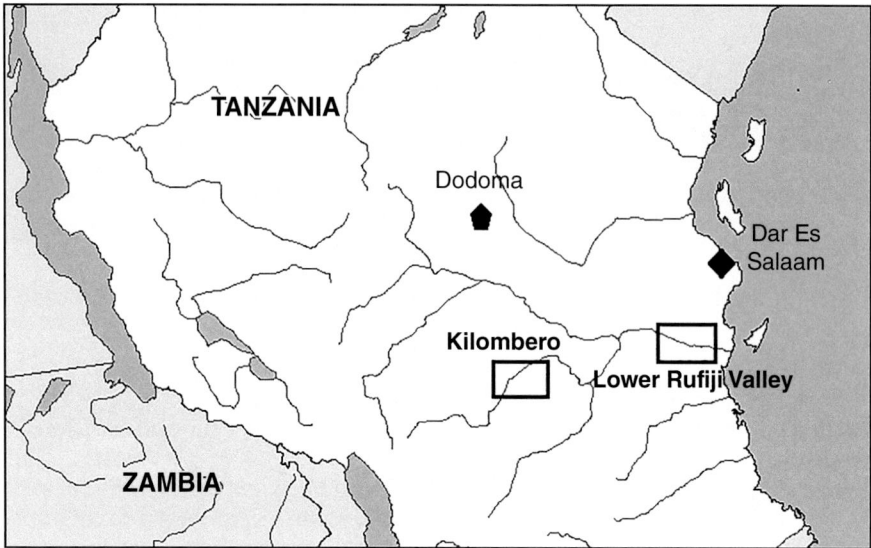

Fig. 3.1. Location of the study areas in the Rufiji River Basin.

3.1.1 Rainfall calendar, flood calendar, and climatic hazards

3.1.1.1 In the lower Rufiji Valley

The chart shows a relatively long rainy season from late October to May and a dry season from June to mid-October (Fig. 3.2). In fact, the slight dip in rainfall in February reveals the presence of a short dry period that the average conceals due to its interannual variability. It is the transition between the seasons of "small rains" (*mvuli*) and "large rains" (*masika)*. These two seasons are used by farmers to carry out two rainfed crop cycles (mainly corn, but also rice flooded by rainwater runoff). The flood calendar depends on upstream rainfall and is shifted in relation to the local rainfall calendar. At the end of the short rainy season, "flash floods" occur, which can have a destructive effect on the most exposed crops. This period is followed by a brief return to low water and then the main flood. The flood allows for two cropping seasons: the flooded rice crop, which was planted at the end of the short rainy season, and the flood recession corn crop. The rice that will be flooded by the rising water is sown in seed holes, with direct drilling, and when it is dry. The beginning of its cycle is therefore rainfed; it depends on the short rainy season. During the flood, the rise in water level must

therefore be compatible with the speed of the rice's growth. In the absence of water control, the rice can be submerged (e.g. as in 2013–2014), and it can also wilt at the early stage due to lack of rainfall or later, flooding (e.g. in 2008–2009). In addition, a flood that is too low favors the development of weeds and disadvantages the fertility reproduction of the non-flooded fields.

Precipitation is also marked by very strong interannual irregularity. While the average rainfall is 867 mm, the rainfall volume has varied between 500 and 1300 mm over the last decades (Fig. 3.3). Moreover, as the lower valley has been the subject of extensive scientific work in hydrology, fairly precise data on the evolution of precipitation are available, with a significant historical perspective. Duvail *et al.* (2014) highlight a clear drop in precipitation over the last 15 years (Fig. 3.3), with a magnitude comparable to that experienced in West Africa during the 1970s and 1980s (see Chapter 1, this volume).

This sharp decline in rainfall over the past 15 years is widely reported by farmers (Tewa, 2014). Since 1999, rainfall has only once exceeded the average established over the previous period (850 mm). This change is characterized by a significant decrease in rainfall volume and the duration of the short rainy season.

When it comes to the flood's evolution over time, there are no long-term records available. It

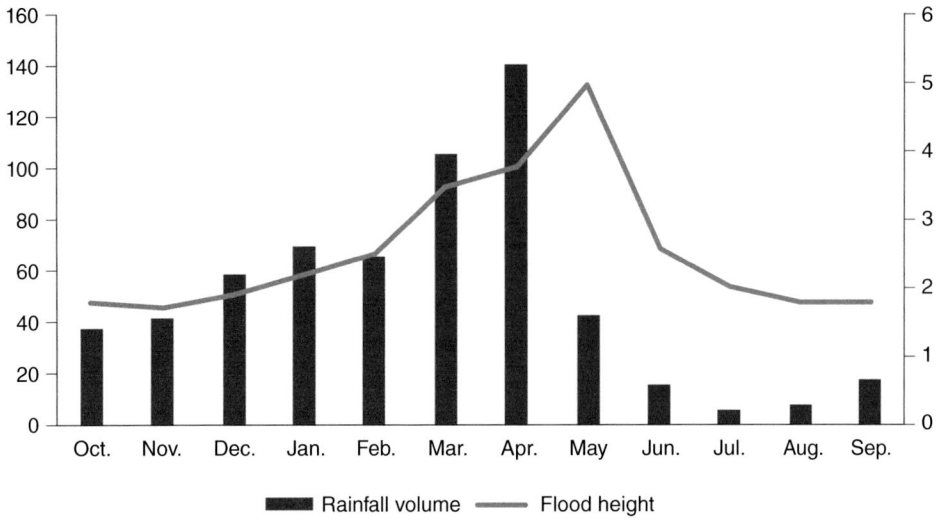

Fig. 3.2. Rainfall and flood calendars. Average 1999–2012; flood: average curve shown for information. (From: precipitation: Utete weather station.)

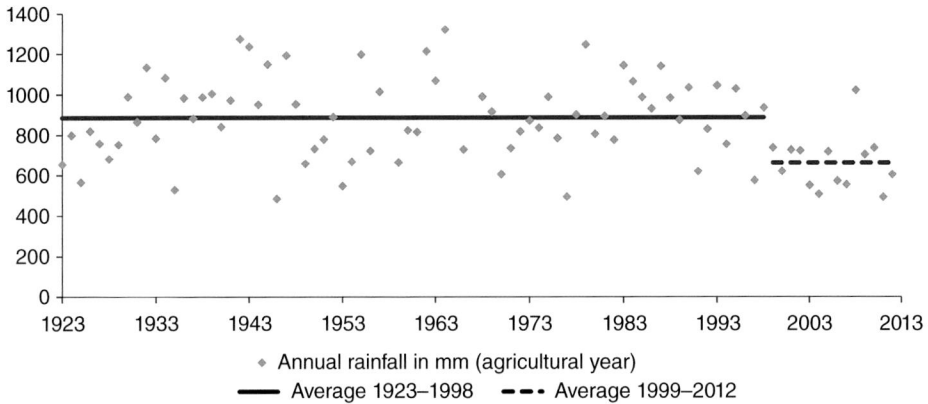

Fig. 3.3. Evolution of annual precipitation volume in Utete and average of two periods (1923–1998 and 1999–2012). (From: Duvail *et al.*, 2014.)

is only in the last 15 years or so that Duvail *et al.* (2014) have undertaken accurate measurements. Over the observations' duration, maximum flood height ranges from 6.5 to 4.2 m, a difference of two meters, which has important consequences for agriculture. Flooding is perceived by the population as a benefit and not a disaster (Duvail and Hamerlynck, 2007). It is in fact the basis for the renewal of the flooded fields' fertility, the possibility of flood recession cultivation freed from the rainfall hazard, and the lakes' recharging with water and fish (main fishing areas).

Hamerlynck *et al.* (2010) attempted to answer the question of vulnerability to these hazards. They define three flood situations:

• The "ideal" situation, where the annual flood is in harmony with the region's agricultural system, the probability of such a situation being 0.25 (i.e. once every 4 years).
• The situation where the annual flood does not cover the totality of the flood plain's fields, a situation whose probability would be 0.60 (i.e. 6 out of 10 years).

- The situation of excessive flooding, which causes significant damage to rainfed and flooded crops but ensures a significant extension of flood recession crops and an increase in fishing potential, with a probability of 0.15. The flood of 2014 belongs to this last category. According to the farmers interviewed, no flood of this magnitude had occurred since the 1997–1998 crop year, attributed to the El Niño phenomenon.

3.1.1.2 In Kilombero

3.1.1.2.1 RAINFALL VOLUME: INTERANNUAL VARIABILITY AND LONG-TERM EVOLUTION. The rainy season here is more compressed and the rains do not really start until December, later than in the lower Rufiji Valley (Fig. 3.4). On the other hand, rainfall volume is much more consistent here and is in the vicinity of 1500 mm (average for the last 20 years).

A long series of daily rainfall records (1927–1972), which we were able to obtain from the former Catholic Mission of Ifakara's experimental station, now closed,[1] as well as more recent series available locally, have made it possible to reconstruct rainfall volume evolution for the last 90 years (Fig. 3.5). It has revealed important

changes and provides a basis for comparison with the testimonies of the people surveyed.

The graph reveals a very high interannual fluctuation in rainfall, ranging from 750 to 2500 mm. Rainfall's great irregularity is an ancient reality with which the populations have always had to deal. Two recent dry periods are clearly visible: the 1970s, which is well linked to the drought recorded at the same time in the Sahel and which is mentioned very frequently by older farmers, and the more recent, but apparently equally marked, period of the late 1990s and early 2000s.

But the most striking information shown by this graph is a much longer and more marked dry period than the two mentioned above, particularly during the years 1940–1955. Farmers' current statements, "There is less and less rain today," must therefore be put into perspective. Their grandparents clearly experienced a worse situation. It would therefore be particularly interesting to identify farmers' responses to these much drier episodes and to understand the reasons why such adaptation seems more difficult for today's farmers.

Finally, the trend in recent years would be for an increase in the average rainfall level, and even more so for an increased frequency of very heavy rainfall, as in 2007, 2014, and 2016.

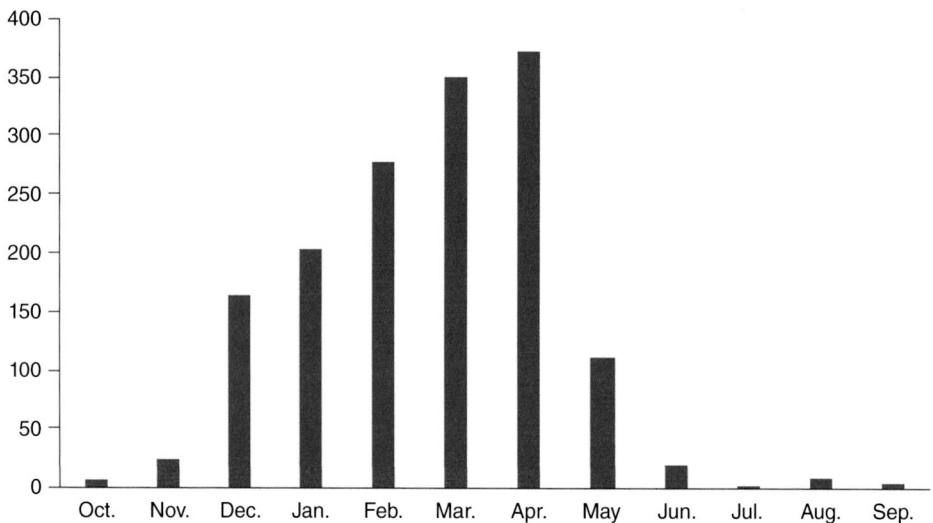

Fig. 3.4. Rainfall distribution in Ifakara, Kilombero Valley (2001–2010). (From: Ifakara Rice Experiment Station and Illovo Plantation in Kidatu; graph: authors.)

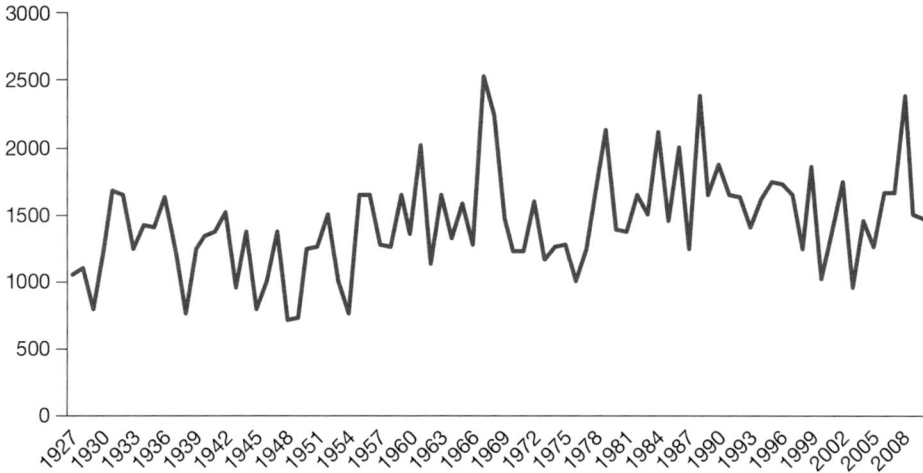

Fig. 3.5. Rainfall trends at Ifakara, Kilombero Valley (1927–2010). (From: Ifakara Catholic Mission 1927–1972 and Rice Research Center 1968–2010 daily climate records; graph: Le Clerc, 2015.)

3.1.1.2.2 SEASONAL RAINFALL DISTRIBUTION: TOWARD A TIGHTENED RAINY SEASON.   We have divided this long-term series of daily records into 20-year periods: 1930–1950 (period 1), 1950–1970 (period 2), 1970–1990 (period 3), and 1990–2010 (period 4, Fig. 3.6). We then compared the bimonthly averages over each of these periods with the most recurrent farmer statements, namely:

- "In the past, we used to see a drop in precipitation in February. This is no longer the case."
- "The rains' arrival has shifted: our parents used to sow in December when the first rains came, today this phase takes place in January."

For the whole 1930–1990 period, February is indeed a time when precipitation ebbs slightly, while over the 1990–2010 period, February precipitation undergoes, on the contrary, a sharp increase (Fig. 3.6). Concerning rainfall's arrival, today (1990–2010 period) we can see a clear delay in the first rainfall compared to the previous 40 years (1950–1990). This situation also characterizes the 1930–1950 period, with annual rainfall that was significantly lower than today. Farmers' perception of a delayed rainy season, the consequent tightening of its duration, and the increase in February rainfall therefore corresponds to reality.

3.1.1.2.3 MORE LARGE-SCALE FLOODS.   The flood calendar is slightly divergent from the rainfall peak. Maximum rainfall is generally reached at the end of April or the beginning of May with a water level of 4–5 m above the low water mark, or even more than one additional meter in years of very heavy rainfall. Given the valley's very flat topography and in the absence of any hydraulic planning, interannual variations in the flooded area and the water level's height are therefore significant. This is a hazard that farmers must manage on the scale of each of their fields.

In contrast to the lower Rufiji Valley (*above*), detailed information on the flooding calendar of the Kilombero and its tributaries is too sparse or too incomplete to be of much use. What remains is the testimony of elderly residents of flood-prone areas since the 1940s stating that no floods comparable to those of 1997 and 2014 (both associated by climate scientists with the El Niño phenomenon) have occurred before (Le Clerc, 2015). More generally, many farmers state, "We are seeing heavy flooding more frequently today."

3.1.2 Environmental heterogeneity

In both the lower Rufiji and Kilombero Valleys, river dynamics have created a very complex environment that no hydraulic planning has come to homogenize. The relative technical and economic performance of these two agrarian systems, and in particular the income generated by the families living there, is based on the skillful farming of this complex and relatively wild environment.

3.1.2.1 In the lower Rufiji Valley

This valley has been shaped by the flooding of the Rufiji River for thousands of years. The present valley and its flood plain are bordered to the north and south by ancient fluvial terraces whose banks dominate the flood plain by one or two tens of meters (Fig. 3.7).

The following different landscape units can be identified:

- The largest, but least valued by farmers, are the ancient fluvial terraces, particularly the one on the left bank (to the north, on the right side of Fig. 3.7). The villages resulting

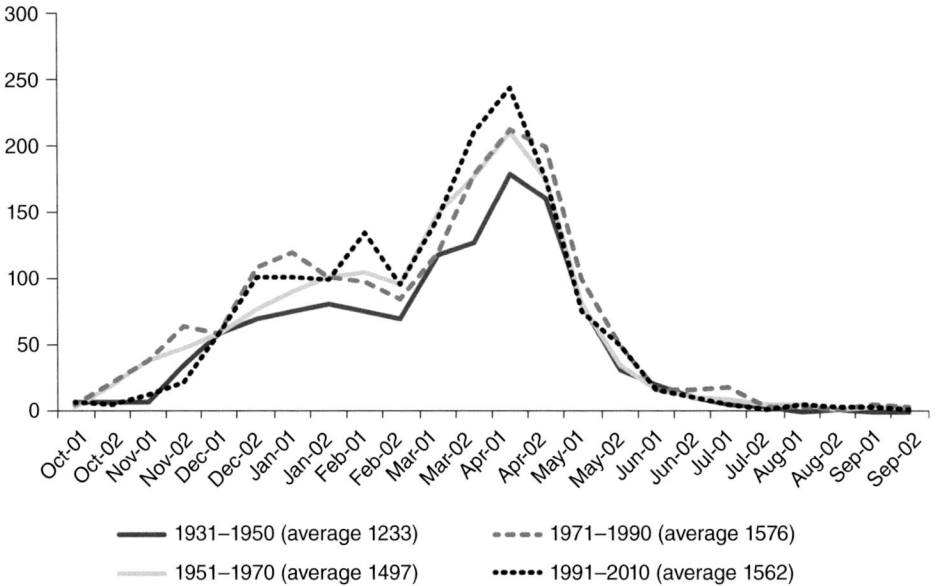

Legend:
- 1931–1950 (average 1233)
- 1951–1970 (average 1497)
- 1971–1990 (average 1576)
- 1991–2010 (average 1562)

Fig. 3.6. Evolution of bimonthly rainfall volumes (1930–2010) by 20-year period. (From: climate data sheets from the Catholic Mission of fakara and the KATRIN research center; graph: Le Clerc, 2015.)

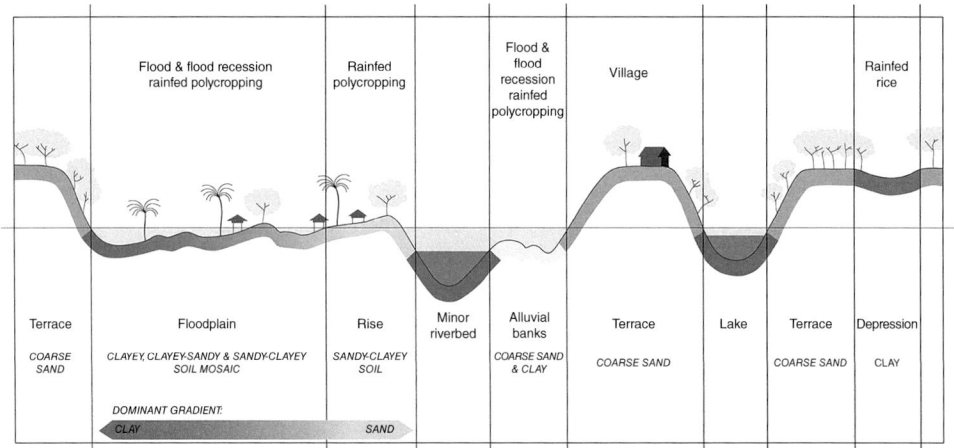

Fig. 3.7. Transect of the Rufiji Valley. (By Jean-Luc Paul.)

from the Ujamaa period villagization (1967–1968) are found here, lined up along the single path that runs along the river. This vast unit is heterogeneous. It is dominated by sandy soils (probably developed on ancient alluvium) and a *miombo*-type tree formation (open forest) exploited by herders (recently arrived and still few in number) and rare hunters. For the villagers, this area is the only source of firewood and timber (carpentry, half-timbering, dugout canoes, pestles, mortars, and furniture). In some places, it includes basins of a few hectares with more clayey soils (*njacha*). These were put into cultivation on the locals' initiative after villagization and are now much less intensively occupied. Villagers criticize them for being able to grow only one crop, rainfed rice flooded by runoff water, in the long rainy season. Moreover, the basins "perched" on ancient fluvial terraces (Fig. 3.7) are not fertilized by the river's flooding. The risk to agricultural production in this area is therefore very high, much higher than in other areas where, as we shall see, two main crops (rice and corn) and three cropping seasons can be combined.

- Lakes occupy fossil riverbeds carved into these terraces. The weakness of their watershed's size makes their episodic or regular recharge by flooding essential (Hamerlynck et al., 2010; Duvail et al., 2014). The lakes make up a unique unit, mostly dedicated to fishing, although the sill separating them from the river can be cultivated in small areas with rainfed crops (corn), during flooding (rice) or flood recession (corn).

- Sandbanks (*tingi/matingi*) occupy the concavities of the meanders of the minor riverbed. They are regularly remodeled by the flood (Fig. 3.7). Although they often have comparable forms (in spindles parallel to the river's main course), they differ in their surface area, height (more or less rounded in shape), age, and therefore in the vegetation that covers them. Exclusively sandy and without vegetation in appearance, they are gradually colonized by herbaceous plant cover. Thus, a reedbed is formed, which, during the flood, slows down the flow and favors the deposit of fine particles carried by the river. Some villagers regularly visit these reedbeds and when they judge that the accumulation is sufficient (at least 30 cm), they decide to cultivate them. Within the same sandbank, especially if large enough, topographical variations create an edaphic patchwork. The very low, clayey areas that are most exposed to flooding are reserved for flood recession corn. The low, clayey areas are suitable for rice cultivation, but corn can be grown in these areas during the short rainy season. The intermediate micro-altitude zones, sheltered from mid-range water levels, are intended for corn cultivation in the short and long rainy seasons. On the bulging tops of some sandbanks, very sandy but almost always sheltered from flooding, long rainy season corn is grown. Here, the presence of mango, cashew, or banana trees testifies to the very low occurrence of flooding; the villagers will therefore build their field hut here (*ndungu/madungu*).

- The river's minor bed is separated from the flood plain by a levee, which remains safe from flooding, except in the case of exceptional floods such as in 1998 during a pronounced El Niño episode. The levee and the sloping bank that connects it to the flood plain are planted with mango and cashew trees (Fig. 3.7). Formerly preferred habitation sites before villagization, these places have been recently reinvested. The soil is rather sandy and especially suitable for sesame cultivation, and sometimes cassava and corn.

- Behind the levee is the flood plain. When the flood is sufficient—when the flow reaches at least 2500 m³/s (Duvail et al., 2014)—the water enters it through a few collapses of the levee and flows through a complex network of channels and marshes and then joins the river downstream (Fig. 3.7). This area is conducive to the sedimentation of the finest particles (the water generally stagnates or flows very slowly) characterized by its much more clayey (and silty) soils and seasonal flooding, which favors rice cultivation. The general trend is that as the elevation decreases away from the levee, the soils become more and more clayey. The more clayey temporary marshes are the site of flooded rice cultivation par excellence.

Flooded areas with lighter soils can support an early rainy season corn planting followed by a flooded rice cycle and eventually a flood recession corn or cotton crop. Where areas have been flooded, corn or sesame can be planted at the beginning of the main rainy season. Finally, the flood plain is an important fishing spot but fishing here is very seasonal.

Despite the fact that this flood plain accumulates the risks associated with flooding and rainy season hazards, the space seems to have long attracted farmers. The lower Rufiji Valley was even known for a long time for its cereal surplus (Hamerlynck *et al.*, 2010). In contrast, the vast spaces of the ancient fluvial terraces on either side, sheltered from flooding, do not seem to have attracted farmers in the long term. Although locally provided with clayey areas that are a priori suitable for rice cultivation, farmers are not able to maintain the soil's fertility, as they have no means of obtaining synthetic fertilizers. Since the time of villagization, several of these areas have been successively cultivated (10–15 years) and then abandoned.

3.1.2.2 In Kilombero

The Kilombero Valley has a different cross-sectional profile. In contrast to the lower Rufiji Valley, the flood plain is not delimited by ancient fluvial terraces with a clear slope. On the other hand, the Kilombero depression is a graben framed by strong reliefs, and its cross-sectional profile is "disturbed" by the presence of alluvial fans. These are formed by the Kilombero's tributaries, both on the left and right banks, and occupy part of the flood plain of the Kilombero itself (Fig. 3.8). The river's water level thus concerns, from a certain level of flooding, not only the flood plain of the Kilombero *stricto-sensu* but also the downstream part of these alluvial fans, this downstream part being subject, on the other hand, to the flooding of the tributary itself (Fig. 3.9). These alluvial fans have levees on both sides of the tributary that makes them up, or even fragments of levees that are evidence of the tributary's former passages (Jatzold and Baum, 1968).

The main landscape units are therefore arranged, on the one hand, according to distance from the river (along a toposequence perpendicular to the valley axis) and, on the other hand, according to position on the alluvial fan. The result is the different landscape units described below, starting from the lowest and most permanently flooded units up to the units that are always unflooded:

- Areas that are flooded each year by the Kilombero River and are permanently covered with a thick layer of water. This vast area, fishermen's domain during the flood, is dominated by a grassy formation on both sides of the river's minor bed (with *Phragmites mauritianus, Pennisetum purpureum, Megathyrsus maximus, Hyparrhenia sp.*)

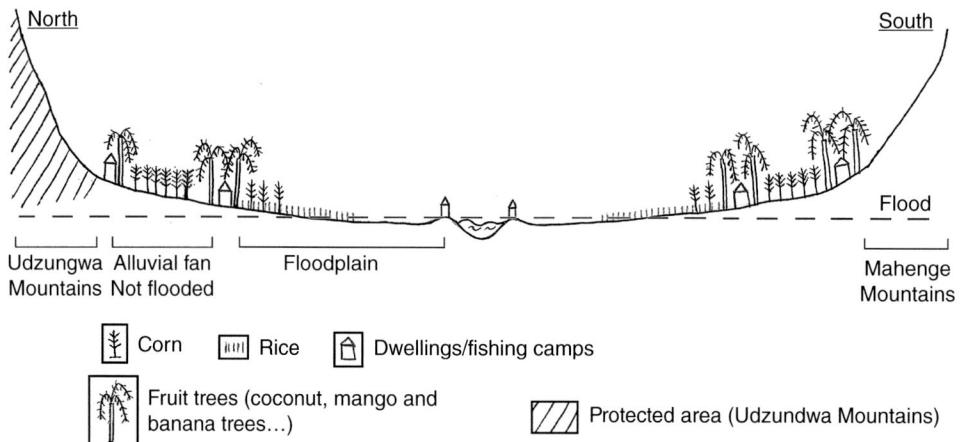

Fig. 3.8. Wide profile of the Kilombero Valley at Ifakara. (From: Le Clerc, 2015, from Jatzold and Baum, 1968.)

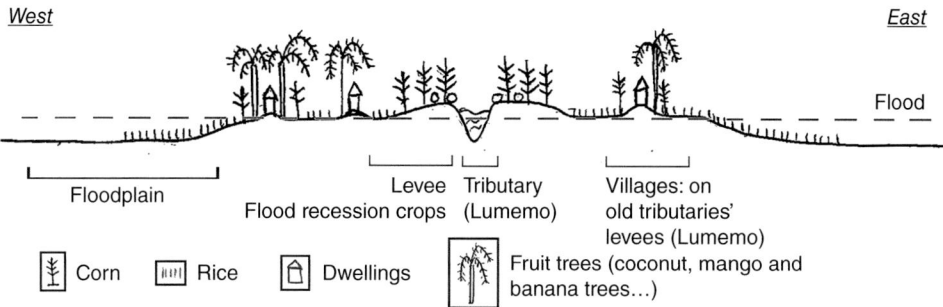

Fig. 3.9. East-west profile of the Ifakara alluvial fan. (From: Le Clerc, 2015, from Jatzold and Baum, 1968.)

and appears very heterogeneous as soon as the flood recedes: it includes (i) small sandy areas, in a high position and corresponding to former levees' remains (some small islands even have mango trees, coconut trees, and a fishing camp); (ii) less sandy areas, a little lower but still relatively high, that emerged quite quickly as the flood receded (some fields are then cleared and corn is direct drilled in seed holes); and (iii) other lower areas, corresponding to channels or basins, more clayey and from which the flood recedes last. The ridges of the previous year's crops are then visible; some of them will be recultivated.

- Grassy plains regularly flooded by the Kilombero but with less water, no trees, and suitable for rice cultivation in a normal year (one rainy season cycle), with flooding usually occurring during the rice boot stage. Between cropping cycles, these areas are covered with low vegetation, the dominant species being *Cyperus distans*, one of rice's main weeds.
- As one moves away from the river, the vegetation gradually reveals a few scattered trees, generally located on mounds corresponding to former termite mounds. Here, too, the flooding, although a little less pronounced, deposits its share of alluvium each year, ensuring the annual renewal of the soil's fertility. This landscape unit merges with the parts of the alluvial fans closest to the Kilombero River. Away from the alluvial fans, this unit is connected to the piedmont plain by large areas of very low slope, rarely flooded (or only for a short time and with a thin layer of water), and dotted with trees.

- There are areas sometimes flooded by the Kilombero's tributaries but not by the Kilombero itself, at least in a normal year (although, again, the lines are rather blurred). These alluvial soils are considered the most fertile in the region. The tributary's proximity, however, implies a very high risk of overflow and flooding in March–April. It is on this land that flood recession cultivation (in the dry season) is possible. Two cropping cycles per year can be carried out there: either two successive cycles of corn (rainfed in December–April and flood recession in June–October) or one cycle of rainfed corn followed by flood recession market gardening (eggplant, okra, squash, cucumber, watermelon).
- The levees (current and former) of the Kilombero tributaries are favored habitat sites, planted with various fruit trees (coconut, oil palm, mango, etc.).
- Finally, the "high" part of the alluvial fans, never flooded, with rather sandy soils, is the junction (break in slope) with the steep slopes of the Kilombero Valley hillsides, itself a very heterogeneous area. Only rainfed corn or sesame cultivation is practiced here.

3.2 Risk Management at the Field Level: Intercropping, Risk Anticipation, and Recovery

3.2.1 The example of the lower Rufiji Valley

In the Rufiji, risk management at the field level has at least two particularities. On the one hand, the dual dependence of cropping systems on rainfall and flooding implies an accumulation of hazards, particularly with regard to rice cultivation. On the other hand, due to the proximity of the immense Selous reserve, the presence of wild macrofauna in the cultivated area is very strong. Faced with these two particularities, the populations of the Rufiji flood plain have long developed specific strategies. However, recent climatic changes are disrupting their effectiveness. For the sake of clarity, we will first describe these long-standing anti-randomization strategies and then analyze how villagers are responding to ongoing changes.

Traditionally, villagers planted short-season corn as soon as the first rains permitted it. Only the vagaries of rainfall affected the crop. Once this short-season corn was harvested, the cropping calendar depended on the area concerned. In flood-prone areas, rice was planted, a crop subject to the dual hydro-climatic hazard (a flood that is poorly synchronized with the rains damages the crop, either by submersion or, on the contrary, by water stress). In areas sheltered from average flooding, a crop of corn or sesame was planted during the long rainy season. In intermediate areas, rice and corn were combined, with the bet being that, depending on that year's particular conditions, one would benefit from the harvest of one crop or the other. Finally, when the floods receded, flood recession corn was planted or, until the 1960s, cotton. By simplifying the cropping calendar to just the main crops (rice and corn), the farmer had the option of four planting periods in the year, as shown in the following diagram (Fig. 3.10):

From a risk management point of view, this staggering over time counterbalanced the restrictions on the dispersion of fields within the different agroecological zones. Indeed, this dispersion is limited by the need for a quasi-continuous

presence in the field to protect the crop from the depredations of wildlife during its entire cycle. The omnipresence of field huts on stilts (*madungu*) in the landscape bears witness to this.

3.2.1.1 The strategy of expanding corn-rice association as an anti-random response to climate change

With the reduction of the short rainy season and the corresponding decrease in rainfall, corn is now planted much later. This leads to a telescoping of the two crops' cycles, and rice is thus interplanted with the unharvested corn. The corn/rice succession has thus become a crop "succession/association" in that the two cycles overlap to a large extent. As a first consequence, although farmers are now opting for shorter-cycle varieties, corn is exposed to flooding (see Photo 3.1). Thus, during the 2013–2014 season, many fields had to be harvested in a hurry, when the water had already risen. A second consequence is that planting rice within the cornfield is more laborious than in an already harvested field.

In addition, where flooding was once considered almost certain, it has become random. As a result, farmers have extended the strategy of combining rainy-season corn with rice to more fields. Here, the rice is sown 1 month after the corn. The two crops grow together in the rainy season. Depending on the year's hydro-climatic conditions, it is hoped that one or the other will provide a significant harvest. Of course, if there is insufficient rainfall and no flooding, or if the flooding is too early and too severe, both crops will be destroyed.

Finally, since the fields are not flat, *staggered* sowing of corn and rice, from the lower parts of the field to the upper parts (the differences in level being limited to a few tens of centimeters),

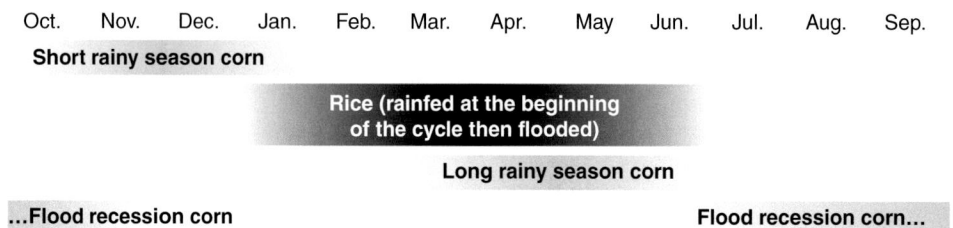

Oct.	Nov.	Dec.	Jan.	Feb.	Mar.	Apr.	May	Jun.	Jul.	Aug.	Sep.

Short rainy season corn

Rice (rainfed at the beginning of the cycle then flooded)

Long rainy season corn

…Flood recession corn   **Flood recession corn…**

Fig. 3.10. Western Rufiji cropping calendar simplified to two main crops. (Courtesy of J.L. Paul.)

constitutes an additional means of increasing the chances of harvesting something, even in the most unfavorable rain and flood conditions.

The 2013–2014 season is a good example of a year in which the deleterious effects of two main hazards combined: an unfavorable rainy season (late arrival of the rains) and an exceptionally high flood. As the rains started very late that year, the planting of corn was itself very late, even though the flooding was greater and earlier than usual. Many cornfields were drowned before the grain matured, while others were harvested in 80 cm or 1 m of water before the flood reached the cobs. In this particularly unfavorable configuration, the late planting of rice may have led to its complete submergence and destruction when the flood suddenly arrived.

3.2.1.2   Catch-up tactics as a response to increased hazards due to climate change

Corn/rice association-succession is also practiced on the sandbank meanders of the minor riverbed (Fig. 3.7). The devastating flood of 2014 also provides a prime opportunity to observe both the considerable risks faced by farmers and the efforts made to limit their scope. That year, almost all of the "short rainy season" corn grown on the sandbanks was submerged by the flood, which also destroyed banana trees, sugarcane, and other minor crops, while the rice crop suffered significant damage due to early submergence.

Under these conditions, the traditional response is to plant a flood recession corn crop. However, this is a *medium-term* catch-up, as the first harvests are only possible from September onward. In order to avoid a period of scarcity, the only possible *short-term* recovery here was based on what remained of the rice fields. At the beginning of May, only the rice plants sown on the slightly higher parts of the fields, the bulging part of the sandbanks, were still emerging. From then on, it was still possible to partially compensate for the losses by taking seedlings from fields that had been spared by the flood and transplanting them, when the floodwaters receded, into the lower parts where the rice had been prematurely drowned (see Photos 3.2 and 3.3). This transplanting technique, which is usually used marginally to increase the area cultivated, has a special nature here. At the cost of

considerable additional work, it allows for a certain catching up of the agricultural season and a notable reduction in the risk of scarcity.

Thus, the cultivation methods of these flood-prone areas—cultivable sandbanks and flood plains—reveal a permanent concern for farmers to limit the risks inherent in the double hazard of rainfall and flooding. Taking advantage of the environment's subplot heterogeneity, combining successions and associations of crops with diverse characteristics, and maximizing the extension of the cropping calendar make it possible to create a diversity of agronomic situations, each corresponding to the anticipation of a particular hydro-climatic scenario. However, these strategies often prove insufficient, as was the case during the 2013–2014 season, and remedial techniques are then mobilized. In all cases, villagers are regularly called upon to make tactical decisions to adapt or even change their initial strategy.

This wise agriculture, which succeeds in making the most of an environment with little human intervention, has aroused the interest of several authors. It has been described as a *Risk Management System* by A. Sandberg (2004). In reference to this author, Hamerlynck *et al.* write, "The Rufiji flood plain farmers have come to grips with the interplay between short rains crops, long rains crops, floods and recession agriculture and the subtle use of the topographical variability, which defines the nature of the soils and their flooding frequency" (Hamerlynck *et al.*, 2010, p. 224).

3.2.2   In Kilombero

The Kilombero Valley and its potential for rice cultivation was the focus of early attention by the colonialists, and later by independent Tanzania's agricultural extension services, and this led to greater integration with commercial trade and the development of tractor and chemical inputs. This is why it is necessary to go back far enough in history to find, in conversation with elderly farmers, cultivation practices that closely resemble those seen in the lower Rufiji Valley.

Because the rainy season here is shorter (*above*), there has never been a true corn/rice succession as in the lower Rufiji Valley. In the past, rice was sown "at Christmas rains," that is, the first heavy rains of December, with the rice being broadcast sown into the corn planted a few weeks earlier.[2] At that time (1950–1970 in Fig. 3.6), the rainy season began a little earlier and conditions were considered more favorable for agriculture. Corn was planted early and harvested well before rice, shortening the hunger gap. The corn-rice association was thus a way of managing the combined effects of the dual hazards of rainfall and flooding.

Our 2015 interviews in this region also show how the joint use of different rice varieties limited the risks involved.[3] The farmer subdivided his land into several plots on which different varieties were sown,[4] making it possible to limit the risks from various sources (climate, flooding, diseases). These varieties were distinguished by:

- A variable cycle length, allowing for adaptation to rainfall and flood fluctuations.
- Late varieties harvested in June, with earlier varieties harvested in March or April. The latter were not very popular, but they were used to get through the hunger gap. They have a distinctive name, grouped under the name "*Msonga*" (*Msonga Meli*, *Msonga Useneguse*).
- "Denser" varieties, intended for beer production.
- Variable resistance to diseases.

In addition to the flexibility allowed by the combination of several varieties, early or late, the harvesting techniques and culinary methods of preparing rice also increased the possibilities of staggering harvests and therefore consumption[5]: daily harvesting of the first panicles that have reached maturity for immediate consumption; harvesting before maturity, at the milky stage and consumption in the form of "*pepeta*" (the grains at the milky stage, unshelled, are fire-roasted and then crushed with a mortar); harvesting just before maturity ("*mchopeko*" rice: the grains are boiled in water and then sun-dried).

Finally, in the flood plain, it was sometimes possible to harvest rice regrowth on the lowest fields in very wet years: the grains that fell at harvest and germinated on the still wet soil allowed for a small complementary harvest. If the flooding was too great, the low yields of the first harvest were then partly compensated by this second harvest (Jatzold and Baum, 1968).

Older farmers unanimously state that the harvests in those days were much better than they are today: two to three acres of crops could feed a family all year long. The granary was never empty or only was a short time before the new harvest.

But these old practices were to be greatly modified, starting in the 1950s, by the emergence of larger, "commercial" rice or cotton farms, 10–20 acres (4–8 ha). The colonial government's aim was to encourage the development of "avant-garde" farmers through easier access to motorization and inputs, in the hope of pulling up neighboring farmers, by transmitting knowledge "over the hedge" (Coulson, 1982).

3.3 Risk Management at the Production System Level: Dispersion, Complementarity, and Risk Mitigation

3.3.1 In the lower Rufiji Valley

At the time of Ujamaa villagization (1967–1968 and 1973–1974), the inhabitants of the Rufiji had to leave their hamlets on the flood plain and gather in villages located on the fluvial terraces. Here, the rainfed agriculture recommended by state services quickly proved unsuitable, and villagers invented a new cropping system, flooded rice cultivation in the terraces' clay depressions (Fig. 3.7). However, depending solely on rainfall, with a range of crops almost entirely limited to rice, and a highly constrained cropping calendar, the villagers found themselves in an increasingly precarious situation. Thus, in the mid-1980s, as soon as government pressure eased, most farmers moved to the flood plain and to the sandbanks of the river. The fertility was better there and above all the range of environmental conditions was greater. In spite of the hazards, and in particular the risk of flooding, the "sand" banks and the areas where the floodwaters spread had not lost any of their attractiveness.

It is known that the combination of different activities by the same family is a guarantee of reduced vulnerability. In the western part of the Rufiji District, this diversification is based primarily on the combination of fishing and farming. Fishing, a male activity, makes up an important part of the income. It provides at least

40% of food needs (by allowing the purchase of corn flour during the hunger gap, cooking oil, salt, etc.). It also covers the vast majority of non-food monetary expenditures (Paul *et al.*, 2012). Agriculture, a mixed activity mainly conducted by women, is essentially food producing, although surplus rice and corn can be sold, and sesame and cashew trees are grown for sale. The first guarantee of security is therefore carrying out fishing and farming activities in parallel, which presupposes the balanced presence of adult men and women in the household. Single women (widowhood, divorce) are always in precarious situations.

But diversification strategies also concern agriculture itself. Access to different landscape "facets" allows, first and foremost, greater food security, but it also makes it possible to obtain cash income that is complementary to that of fishing. However, several factors stand in the way of this access. The first is, due to the proximity of the Selous Reserve, the strong presence of macrofauna (warthogs, baboons, antelopes, passerines, and even hippopotami and elephants). During the entire cultivation cycle, from sowing to harvesting, day and night surveillance of the field is essential. This explains the scattering of the landscape with field huts on stilts (*madungu*). The wife and young children live there during the cultivation season. In addition, the cultivation of a field is never an isolated act. To keep unwanted wildlife away, one always makes sure that the neighborhood itself will be cultivated by relatives, associates, or friends. Thus, if for a given cropping season, a household tries to cultivate several terroirs (sandbanks, flood plains, or clay depressions of fluvial terraces), the household will favor the surveillance of one of them and expose the other fields to a high risk of wildlife depredation. On the other hand, the change of terroir from one season to the next, during the same year, is frequent and does not lead to this risk exposure.

The second factor limiting access to diversified terroirs is that some of them are small. While a significant land reserve still exists in the flood plain, the levee and its sloping bank are small. However, it is only here that cashew cultivation is possible,[6] the only cash crop of any importance since the near disappearance of cotton after villagization. Villagers who have fields on the levee generally have direct access to the flood plain to the south and, in some places, to the minor riverbed's sandbanks to the north. The sandbanks also have a limited extension. Their main characteristic is high flood exposure; this is an advantage during a flood too small in magnitude to completely penetrate the flood plain. It is also a disadvantage in the case of a large and rapid flood. Households forced to cultivate only this land are therefore very vulnerable, while those who also cultivate another land find it a very useful complement.

Agricultural diversification strategies (in addition to the exploitation of subplot diversity mentioned in section 3.2.1) are therefore based on the successive rather than synchronous development of different facets during the year. The question of access to the different terroirs is therefore crucial. From a legal point of view, access to the flood plain is relatively easy, as long as it is not monopolized by industrial agriculture projects. However, the households' demographic structure sometimes opposes this access. Elderly households, single-parent families, families whose husbands are almost exclusively involved in fishing, and small families with no strong family ties are not very mobile. They are often forced to limit themselves to farming land that is easily accessible from the village, namely the clay depressions of the terraces and the river's sandbanks. Their situation is more precarious. Finally, a few families who arrived during the villagization process and who have a rainfed agricultural tradition essentially exploit the levee's non-floodable sloping bank. Their dependence on rainfall conditions also places them in a precarious situation. As an example, the Table 3.1 characterizes some agricultural situations based on Tewa's work (2014); this table is not exhaustive.

3.3.2   In the Kilombero Valley

In this region, the villagization of the 1970s did not prevent access to diverse ecosystems and thus did not have deleterious effects as in the lower Rufiji Valley (*above*) or the Iringa region (Chapter 2, this volume). Although settlement in the Kilombero Valley was originally dispersed, with families settled on small mounds protected from flooding, there had been several moves to cluster dwellings on the levees of the alluvial

Table 3.1. Characterization of some agricultural situations in Western Rufiji District. From Tewa (2014).

Access to the terroirs	Characterization	Agricultural income per worker per year (Euros, 2014)
Levee and its sloping bank Flood plain Possibly sandbanks	Thanks to the diversification possibilities: – Saturation of the work calendar – Good labor productivity – Risk reduction	800–900
Sandbanks only	High exposure to hydrological risk Diversification limited to subplot strategies	250
Levee and its sloping bank	High exposure to climate risk Diversification limited to rainfed crops Possibility of cashew tree cultivation	500
Clay depressions Sandbanks (in flood recession)	High exposure to climate risk Dependence on flooding for flood recession crops	500

fans in the past, so that settlement was already fairly well clustered.

In this region, this period also corresponds to the construction of the Dar es Salaam-Lusaka railroad. At the end of construction, groups of workers who had worked on the construction site settled permanently in the region, participating in its demographic growth, the emergence of new villages along the railroad, and the establishment of a class of landless workers. This last point is not without importance for the subsequent evolution of the region's production systems.

3.3.2.1 *The importance of the corn/rice combination at the farm level*

Until the 1960s, corn was only planted in association with rice at low density. Its function was to bridge the hunger gap "while waiting for the rice harvest" (*supra*). In the following decades, corn became more important and was grown on its own fields. Indeed, as rice was sown later and later (*below*), and as the use of tractors for soil preparation became more widespread, it became more difficult to continue with the crop combination (with earlier sowing of corn, as before). This made it easier to plant corn on separate fields, especially since the possible lengthening of the hunger gap, linked to the delay in planting rice, reinforced corn's role as a hunger crop. Gradually, all farmers began to grow rice and corn on separate fields, with corn increasingly grown on the still-excavated fields located higher up in the valley's toposequence. This development of corn

cultivation was also encouraged by the National Maize Plan, promoted from the mid-1970s onward (Chapter 2, this volume), although the Kilombero Valley was not one of the main target areas.

Recent history is thus marked by the decline of corn-rice association in favor of cultivation on separate fields within the farm. Rice occupies most of the land, while one or two acres are reserved for corn, mainly for self-consumption. In the farmer's minds, the main crop is rice, with corn remaining for the hunger gap season (March–June).

3.3.2.2 *Dispersing rice fields to reduce risk*

Given high rainfall variability, particularly at the beginning of the rainy season, and the unpredictability of both the water level in March/April and the flood's duration, cultivating several sites makes it possible, as in the lower Rufiji Valley, to divide risk and ensure a minimum production level each year. Farmers therefore systematically try to have several rice fields located on sites with different exposure to flooding. This strategy is, however, limited by the farmer's ability to finance the crop year (*below*) and made increasingly difficult to implement with increased land pressure, the relative scarcity of land, and its increased distance from dwellings. As a result, young farmers who have recently settled down sometimes have to make do with a single field, exposing themselves more directly to the impact of annual fluctuations in climate and flooding.

3.4 Further Integration into Market Exchanges and "Agricultural Modernization" in the Kilombero Valley: Toward New Hazards?

3.4.1 Integration with market exchanges and reduction of varietal diversity

The joint use of early and late varieties and the staggering of each cultivation operation (sowing, weeding, harvesting) had a threefold objective: avoiding excessive workload peaks, adapting to fluctuating climatic conditions and the unpredictable level of flooding, and thus shortening the hunger gap period as much as possible.

Today, some farmers still mention this staggering of sowing from mid-December to early February, but entry into the market economy has led to a significant reduction in agrobiodiversity. Indeed, as Kilombero has become an important rice-growing area, the varieties chosen by farmers are in fact those that are most valuable on the market and for which Kilombero is known: highly aromatic but long-cycle varieties. The length of the cultivation cycles therefore tends to be homogenized around 5–6 months. Thus, the range of cultivated varieties is now reduced to about ten. A poor harvest can therefore no longer be compensated for by using short-cycle varieties to limit the hunger gap. The importance of corn and its development during the 1980s is therefore easy to understand.

However, farmers remain attached to their traditional varieties, especially since they remain highly adapted to the conditions of the valley and are still highly prized in urban markets.[7] Although sensitive to lodging, these high-straw varieties (1 m high) allow better adaptation to subplot heterogeneity and the changing height of the water table. Their diversity also makes it possible to spread out workload peaks a little more.

3.4.1.1 The difficult financing of the agricultural year

Today, farmers' cash flow is often partly devoted to covering rice cultivation costs: almost all farmers use a tractor for plowing and harrowing. Even on small farms, the farmer always uses the tractor in December for pre-planting plowing, and then for post-sowing harrowing. Each of these operations is done at a cost of TZS 40,000 per acre. In addition, larger farms will need to pay for the day workers required for manual weeding (or purchase of herbicide), harvesting, and, if necessary, transport. As far as inputs are concerned, only one herbicide is commonly used for rice, namely 2,4D (2,4-dichlorophenoxyacetic acid).[8]

In the absence of favorable access to credit, it is the sale of paddy that covers the season's costs. Therefore, the previous year's paddy stock determines the following area planted, limited of course by the family labor force's maximum capacity (generally set at 2 acres/worker). A year with a good harvest and high selling prices allows a farmer to cultivate all their fields the following year. A poor harvest (exceptional flood), unforeseen expenses (e.g. illness), or a drop in prices will directly limit the capacity to plant. Thus, family farms' cultivated area fluctuates from year to year according to the combined fluctuations of climate, prices, and family contingencies.

3.4.1.2 Shifting the sowing date: climatic disruption or increased market dependence?

Today, producers all sow rice in January, whereas older people systematically mention sowing in December "at the time of the first rains" (*supra*). Is this delay in sowing rice solely the consequence of the rains' general delay over the last few decades, or is it also a consequence of the increased use of inputs and tractor services in the context of deficient financing of the agricultural year?

Since crop advances are usually financed by selling the previous year's paddy crop as needed, attempts are made to sell paddy at the best price possible. Since December is a period of rising prices, farmers postpone the sale of their paddy as much as possible, even if it means delaying the sowing of rice too long and exposing it too early to flooding. The search for this difficult balance between the best paddy selling price and the sowing date helps explain why rice is now sown in January, much later than in the 1960s or 1970s, to be harvested from May to July.

Thus, it seems that the delay in sowing toward January is as much the consequence of a delay in rainfall, as often presented by valley farmers, as it is an effect of the difficulties in financing the crop, particularly the tractor.[9] This example illustrates the difficulties encountered when discussing climate change and its consequences with farmers, and the precautions to be taken in this regard: "talking about the climate without talking about it" (*see "Introduction"*) and being careful to place the possible consequences of a lasting change in the climate in the context of those resulting from changes of another nature.

3.4.2 Increased socioeconomic differentiation

The region's increasing monetization, the decline in the relative prices of cereals and labor (linked in particular to the implementation of the structural adjustment plan), increasing access to tractors for soil tilling, and the presence of a salaried labor force (day laborers) are all factors that have favored an increase in cultivated areas and production for some families. But the widening of the cultivated area relies on a daily paid labor force with limited access to land.

Moreover, once corn and rice were grown on separate fields, and with an almost synchronous cropping calendar, the two crops became competitors for labor. Combining them beyond an area of a few acres results in significant labor peaks. As a result, growers now almost systematically use off-farm labor force for weeding and harvesting.[10] Access to a low-cost labor force then allows a category of farmers—those better-off—to expand the cultivated area, especially after the arrival of private contractors offering tractor plowing services, which also facilitated this expansion. Very quickly, from the 1990s, families with the means to expand their areas were able to produce about four times their annual consumption (Kato, 2007).

The economic results of the main types of farms encountered in the region were estimated by Le Clerc (2015). By way of example, we will describe the following cases:

- The most common production system in the region is the one implemented by two working family farms cultivating between 2.5 and 6 acres (1–2.5 ha). The land is divided between flooded rice in the valley lowlands (clay zones) and rainy-season corn on the alluvial fan's flooded lands.

These farms, which are often based on the old levees (with a garden orchard), usually fish almost daily, which provides a significant part of the income. While corn is planted on one to two acres for self-consumption, without inputs or outside labor, the rest of the area is planted with rice (2–5 acres), using the tractor for plowing and harrowing. The anti-dicotyledons herbicide 2,4D is often used, but chemical fertilization is not, as it is considered too expensive. As with all small-scale family farms, rice is sold as the year progresses. The rice stock is usually exhausted by January, with most of the production sold before prices reach their peak. Around the houses, there are usually coconut, banana, and oil palm trees, and a small vegetable garden, which provide a regular income throughout the year. The income generated by this type of farm (including self-consumption) is very low, in the order of 220–370 euros/laborer/year.[11] Extreme weather events are therefore a permanent threat and a brake on these small family farms' economic development insofar as, without any alternative financing, their production capacities are based solely on the previous year's performance.

- Farmers with access to fields close to the river's minor bed and suitable for early dry-season market gardening, once the flood has receded, fare better. Although these fields are very fertile and cannot be used for rice in the rainy season, the water level being too high in April, they are very suitable for off-season market gardening. The water is brought thanks to a bucket or watering can. In addition to the 2.5 and 4 acres (1–1.6 ha) of flooded rice in the clay areas and 1–2 acres of rainy season corn, vegetables are grown on 0.5 acres (this is the maximum allowed by the labor load required for hand watering). The financial resources from market gardening make it easier to call on off-farm labor force in a timely manner with the goal of more and higher quality production. The income generated can then reach the equivalent of 700–900 euros/laborer/year.

- There are also slightly larger farms, with 5–7 acres, 3–5 of which are flooded rice (4 t/ha) and two of which are corn (4.5 t/ha), using chemical inputs and employing many day workers. Yields are higher and farm income exceeds 1300 euros/laborer/year.

• Finally, there are large motorized patronal farms (a secondhand tractor and a tractor driver) with a larger surface area, between 50 and 70 acres. With the exception of tillage, the entire crop management sequence is carried out manually by external workforce, with weed killer (2.4D), but without recourse to chemical fertilizers. Yields are therefore similar to those of the small family farm, and the annual value-added does not exceed 400 euros/ha. The tractor is mainly made profitable as a service provision to other farmers, as demand is very high in the Kilombero Valley.

3.5 Large-Scale Development Projects and "Intensive" Rice Cultivation: Reducing Hazard and Vulnerability?

3.5.1 In the lower valley of the Rufiji River

Since the beginning of colonization, the lower Rufiji River Valley, like other regions with comparable characteristics, has been the object of ambitious planning projects based on the erection of large dams with the combined functions of electricity production, flood control, and irrigation of the downstream part of the valley (Duvail et al., 2010). The recent arrival of new foreign investors and legislative changes facilitating their reception are now reviving these large projects, which are presented by their promoters as a privileged path toward adaptation to climate change (National Irrigation Master Plan, NIMP, 2002). By greatly reducing the risk of flooding through extensive artificialization of the environment, they would be able to substantially increase agricultural production while securing it in the long term. The massive creation of jobs for local populations and the consequent improvement of their standard of living are also put forward as benefits of such projects.

In Rufiji, about 20 investors, mostly foreign, are reportedly currently in the pipeline (Tewa, 2014). Specifically, in the study area, the project is expected to result in the creation of an irrigated rice farm covering more than 8000 ha of flood plain. In order to provide investors with long-term land security, the State has previously asked village authorities to validate a zoning of the communal territory that distinguishes between the area currently under cultivation and a "land reserve." It is this "land reserve" that the Tanzanian government intends to lease to national or foreign investors, if possible, with the agreement of the plenary village assembly (or otherwise in an authoritarian manner as authorized by law).

As we have written elsewhere (Cochet, 2014), most of the investment projects proposed in the countries of the South concern areas that benefit from conditions that are eminently favorable to agriculture (soil fertility, water resources, accessibility). Therefore, most often, those areas have long been occupied by agricultural societies characterized by a relatively high population density and labor-intensive production systems. Some of these characteristics are found in the lower Rufiji Valley, but also in several large African deltas where both agricultural and livestock activities are concentrated: the inland deltas of the Niger (Mali), Awash (Ethiopia), and Chari (Cameroon/Chad) and the Tana Delta (Kenya). The large-scale investments made by foreign public or private agents actually result in the substitution of pre-existing agrarian systems by new ones. Those investments are not a conquest of "virgin" land (i.e. unexploited land for which the opportunity cost of land, water resources, and labor would be zero), as it is often incorrectly stated. Real progress in terms of production, value-added, and job creation is therefore uncertain and must be analyzed on a case-by-case basis (Ibidem).

A thorough reflection on projects concerning the lower Rufiji Valley, in terms of economic evaluation, is not possible within the framework of this chapter, especially since the fieldwork carried out within the context of this research program was not oriented in this direction. However, the knowledge acquired on the dynamics of the agrarian system that would be affected by this type of project allows us to advance that the central problem with creating a large dam is the suppression of flooding. This would mean no more fishing, dried-up lakes, and the end of flooded and flood recession agriculture. In the case of the lower Rufiji Valley, however, we have seen that families' agricultural income depends closely on their ability to combine different cropping systems that take advantage of the environment's different "facets." Among these different areas, the flood plain is the most sought after by farmers because of its greater potential, despite the risks involved. However, it is precisely this space that is today coveted by development projects and that risks completely eluding the region's inhabitants, or even becoming the scene of a massive eviction.

Despite compensation from the Tanzanian government (in what amount?)[12] and possible salaries distributed by the project (the number of which is generally greatly overestimated at the time of the project's announcement), such a project is likely to result in an increase in populations' vulnerability and precariousness, either by depriving them of access to one of the areas exploited or by reducing all possibilities of compensation and managing the environment's heterogeneity.

In addition, beyond these projects' often very high investment costs, they are supposed to generate a higher value-added per surface area unit than "traditional" production systems. It remains to be seen over time, as does the planned structures' capacity to withstand very large floods, the frequency and intensity of which, as we have seen, are likely to increase.

The paradox of such planning, whose objective is to reduce the risks incurred by climatic hazards, through artificializing and homogenizing the environment (in the hope of increasing yields and reducing their variability), is precisely that there is a high probability that they will increase populations' vulnerability.

3.5.2 In Kilombero

In order to facilitate foreign direct investment (FDI), SAGCOT was created in 2010 to develop agricultural production in a corridor linking Dar es Salaam to Zambia. In the form of a public-private partnership, SAGCOT is the first initiative in a series of corridors on which large-scale agricultural production development programs will be focused to "ensure the country's food security, reduce poverty, and increase resilience to climate change." In a now familiar pattern, it is proposed to initially promote the establishment

of large, capital-intensive, high-yielding farms, in the hope that this economic development will lead to easier access to inputs, credit, moto-mechanization services, markets, and training for the surrounding small farms, which would then see their yields increase.

Within the Kilombero "cluster," three sites have been identified for large-scale, privately owned farms, two of which, each with an area of approximately 5200 ha, would be for rice cultivation: the Ngalimila and Kihansi sites in the southwestern part of the valley (SAGCOT, 2012).

We will not address here the question of these spaces' current occupation by valley populations and the possible consequences, in terms of social eviction, of the implementation of these two projects. However, the analysis will focus on the support program that would be put in place to ensure that innovation would spread to small producers.

To address this question, we looked at rice company Kilombero Plantations Limited (KPL), which has been operating since 2008 near the town of Mngeta, in the southwestern part of the study area for this research program. KPL has a 5000 ha concession for 99 years. Since 2010, KPL, in partnership with USAID, has been implementing an agricultural extension program for farmers in the surrounding area aimed at increasing rice production through the use of selected seeds, inputs, and a specific crop management sequence.

We drew inspiration from the evaluation methods commonly used in the economic evaluation of projects. They are based on the measurement of a differential between the situation resulting from the project's implementation, on the one hand, and the situation that would have prevailed if the project had not been implemented (counterfactual or "without project" scenario), on the other. In the absence of a diagnosis of the pre-project situation in the KPL project's actual location, the agrarian system described in the preceding pages was chosen as the "reference situation." Insofar as this is a region close to the one concerned by the project, from a biophysical point of view (geomorphology, pedoclimatic and hydrological conditions), but also from a socio-economic perspective (population's origin and demographic density, cultivation and livestock techniques, access to markets, etc.), this choice seems reasonable and likely to limit the biases inherent in this type of comparison.

KPL offers agricultural training in the System of Rice Intensification (SRI). This technical package includes selected short-strawed varieties (SARO5), row seeding, and fertilization (NPK and urea). Depending on the fields' topographical position, it is possible to use either direct drilling (e.g. rainfed rice in unflooded areas close to the mountains) or rice paddy fields with nurseries for flooded areas. After each training session, farmers leave with the necessary tools to implement this technical package at home, on a small area. In 5 years, more than 7400 farmers have benefited from this training.[13]

Interviews conducted by Le Clerc with farmers using this new technical package revealed the following observations:

- Fields cultivated with SRI produce significantly higher yields (up to 7 t/ha of paddy), and the income from the sale of paddy helps to repay the SRI loans. In addition, farmers who implement this crop management sequence on one of their fields also use synthetic fertilizers or pesticides on the other rice and corn fields on their farm, thus obtaining better yields. Thus, farmers who benefit from this program do not limit their use of inputs to SRI fields.
- Yet, paradoxically, the adoption of the SRI technical package is very limited on any given farm. Indeed, small family farms apply this technical package to only one acre, while owner-managed farms apply it to a maximum of 20% of their land. The rest is cultivated according to the "usual" techniques.

Why do farmers devote such limited area to this new crop management sequence, even though it is effective in terms of yield increase? Several reasons can be given:

- Row seeding or, a fortiori, transplanting, requires far too much time to be generalized to all fields. For the farmers interviewed, the implementation of these practices results in a significant increase in labor and input costs that can only be done in small areas.
- The valley's traditional varieties sell much better and it is often difficult to sell the production from selected seeds that are not very aromatic (SARO5).

- This crop management sequence requires high financial resources per unit area (labor, inputs) and recourse to credit is unavoidable.
- On the other hand, although these systems provide at least partial protection against climatic hazards (especially for transplanted fields), the financial risk seems disproportionate and can make the farmer even more vulnerable in the case of crop failure. This is a crucial issue in Kilombero.

In the region studied, we saw that for a majority of farms, especially the smallest, the area actually cultivated depends on the previous year's harvest. It is thus conceivable that the overall budget for a year of SRI cultivation is so high that a poor harvest would be disastrous for households if all the available land were cultivated that way. Dedicating part of the land to the SRI system and the other part to the "traditional" system makes it possible to hope for a better harvest in a normal year, without incurring excessive costs. This is a way to increase income while limiting the risk of climatic hazards. Given the climate's great variability, farmers' ability to withstand a year of low yield seems to be at the heart of their reasoning concerning their cropping practices.

Here, as in many parts of the world, the number one objective for public authorities, companies, and the accompanying projects is always to increase yields, presented as an end in itself. However, for a family farm, the objective is not to achieve the best yields because such a strategy is always linked to high agronomic and economic risks. The first objective is to ensure an income that will allow the household to live and, if possible, to invest. Thus, the family farm is characterized by a permanent concern for limiting the risks of obtaining an income incompatible with the family's survival.

In the case of the program implemented by KPL, the crop management sequence that was popularized and supposed to increase resilience to climate change, in accordance with SAG-COT's recommendations, actually increases the risk incurred by farmers. It is this increased risk that is the main limitation to disseminating the crop management sequence.

Farmers' frequent use of the inputs provided by the project beyond the field covered by the recommended crop management sequence is also instructive. It is often proof that farmers have a great need for the inputs offered to them, but not always for the techniques associated with them. It illustrates once again that the promotion of an overly rigid, one-size-fits-all technical package can be counterproductive. Indeed, a technique that is not adapted to farmers' constraints (e.g. row sowing, which requires too much work) is enough to block the entire dissemination process, and in particular the use of new inputs, unless they are "diverted" to other uses, which are considered more productive.

3.6 Conclusion

In the Kilombero and Rufiji flood plain valleys of southern Tanzania, the agrarian systems' performance and the income generated by households are based on the skillful exploitation of a complex and rather unartificial environment. Here, risk mitigation strategies cannot be assessed without a detailed knowledge of the environment's micro-heterogeneities and the practices implemented to take advantage of them, particularly at the field level. The combined corn/rice crop on the sandbanks and in the flood plain of the lower Rufiji Valley offers an eloquent example of this fine and evolving adaptation to rainfall and flooding hazards. Here, intercropping is not only a "strategy" for anticipating and reducing risk, which makes it possible to "not put all one's eggs in one basket," or a practice for limiting sanitary risks. It is also a means of adaptation and a path toward management that takes the characteristics of the beginning of the rainy season into account (date of the first significant rainfall, spacing of the first rains, rainfall volume, etc.).[14]

Other examples could illustrate this point. For example, Crane *et al.* (2011) describe the adaptation of practices at the cropping system level in a region of southern Mali: on very heterogeneous soils (sandy/clayey) and taking the rainfall hazard into account, farmers sow sorghum and millet together in the same field. After plants have emerged and in view of the seasons' first rains and the soil moisture content, farmers keep only the sorghum plants if the season looks good, or on the contrary, only the millet plants, if the rains seem insufficient.

On the other hand, an approach in terms of production system (and activity system) allows us to understand how access to different ecosystems and the combination of different cropping systems and off-farm systems (such as fishing, in this case), limit risks, this time at the family level, by relying as much as possible on the complementarities offered by the range of activities accessible to farmers. The families in the best position, generating the highest income and better equipped to face the various hazards they are confronted with, are always those that can combine the greatest number of crops and extra-agricultural activities in a variety of spaces or landscapes. This combination is of course dependent on the availability of means of production (tools, livestock, inputs, etc.), labor, and market access conditions.

The hazard is not only climatic—far from it. The difficulties Kilombero Valley farmers encounter in financing their rice crop have shown us this. Here again, there are countless examples that could reinforce this point. In the case of the Masa-Bugudum villagers of northern Cameroon, Jean Wencélius (2016) demonstrates that difficulties in accessing production factors like seeds and plows (the inability to obtain them at the right time), the loss of a laborer on which one was counting (displacement, social obligations), or the morbidity and fatigue inherent to the hunger gap, all play a role in the failure to carry out certain cropping projects. These factors are much more important than the climatic hazards that farmers have ultimately learned to integrate into their strategies or to compensate for in one way or another. For them, climatic uncertainty is a certainty; the "real" hazards, those that are "unpredictable," come from elsewhere.

Notes

1 The exploitation of such an unpublished series by non-specialists is delicate and would deserve some preliminary tests (e.g. the Pettitt test to check the series' homogeneity, as advised by G. Beltrando, personal communication January 13, 2016).

2 Small vegetable gardens occupied the raised mounds sheltered from the flood; squash, eggplant, okra, African eggplant, and cassava were grown.

[3] In Rufiji, about ten varieties of rice are used. They differ in cycle length (3–6 months), height of booting stage, organoleptic qualities, etc. It is known that with the current climatic changes, Rufiji farmers have abandoned long-cycle varieties in favor of medium- and short-cycle varieties. It is likely that in-depth historical investigations would reveal that here, as in the Kilombero Valley, the use of this range of varieties corresponded to anti-risk strategies.

[4] The first varietal inventories date from 1903: a German captain sent 26 cultivars to the colonial government to test their characteristics. It emerged from this study that a long local selection process had allowed the development of an important agroecological diversity (Stuhlmann, 1909).

[5] Similar harvesting and cooking techniques are found in the Rufiji.

[6] Further east, rainfall allows cashew cultivation on the fluvial terraces.

[7] Much more, for example, than the medium-sized and semi-aromatic varieties promoted by research centers, such as SARO5.

[8] Very few farmers use fertilizer except in low fertility areas, as flooding still plays a major role in the reproduction of fertility. Corn, intended for self-consumption (especially during the hunger gap), is generally cultivated by family labor and without inputs.

[9] In addition, the still insufficient tractor availability also plays a certain role in delayed sowing, especially since the fields are difficult to access. The tractor owner gives priority to the largest and best-placed fields.

[10] A harvest results in an overripe, more brittle rice, which sells for less at the huller.

[11] This is hardly more than in the base Rufiji Valley, where farmers have no access to tractors and use almost no "modern" inputs.

[12] As is the case in many regions of the world, eviction compensation is likely to be based on a significant underestimation of the value-added produced by farmers in these areas, an underestimation based on a lack of knowledge of local practices, as well as the failure to take into account associated crops and self-consumption when calculating income.

[13] According to the program manager at KPL in Mngeta, interviewed on May 15, 2015 by Le Clerc.

[14] We thus agree with Richard, who wrote about intercropping as a risk management practice: "They are looking for the combinatorial logic in intercropping where what matters to (farmers) is sequential adjustment to unpredictable conditions" (Richard, 1993, p. 67).

References

Cochet, H. (2014) Accaparements fonciers et grands projets agricoles privés: exclusions paysannes ou création d'emploi? In: Boussard, J.M., Cochet, H., Coste, J., Delevoye, J.-P., Dumazert, P. *et al.* (sous la dir.) *Les Exclusions paysannes: Quels impacts sur le marché international du travai*. AFD, Conférences et Séminaires, Paris, France, pp. 53–62.

Coulson, A. (1982) *Tanzania: A Political Economy*. Oxford University Press, Oxford, UK, 348 pp.

Crane, T.A., Roncoli, C. and Hoogenboom, G. (2011) Adaptation to climate change and climate variability: the importance of understanding agriculture as performance. *NJAS — Wageningen Journal of Life Sciences* 57(2011), 179–185.

Duvail, S. and Hamerlynck, O. (2007) The Rufiji River flood: plague or blessing? *International Journal of Biometeorology* 52(1), 33–42.

Duvail, S., Médard, C. and Paul, J.-L. (2010) Les communautés locales face aux grands projets d'aménagement des zones humides côtières en Afrique de l'Est. *Politique Africaine* 117, 149–172.

Duvail, S., Mwakalinga, A.B., Eijkelenburg, A., Hamerlynck, O., Kindinda, K. and Majule, A. (2014) Jointly thinking the post-dam future: exchange of local and scientific knowledge on the lakes of the Lower Rufiji, Tanzania. *Hydrological Sciences Journal* 59(3–4), 713–730.

Hamerlynck, O., Duvail, S., Hoag, H., Yanda, P. and Paul, J.-L. (2010) The large-scale irrigation potential of the lower Rufiji floodplain: reality or persistent myth? In: Calas, B. and Mumma Martinon, C.A. (eds) *Shared Waters, Shared opportunities: Hydropolitics in East Africa*. IFRA/Mkuki Na Nyota, Nairobi, Kenya, pp. 219–234.

Jatzold, R. and Baum, E. (1968) *The Kilombero Valley, Characteristic Features of Economic Geography of a Semihumid East African Floodplain and Its Margins*. Weltforum Verlag Munchen, Dillingen, Germany, 154 pp.

Kato, F. (2007) Development of a major rice cultivation area in the Kilombero valley, Tanzania. *African Study Monographs* Suppl. 36, 3–18.

Kilembe, C., Thomas, T.S., Waithaka, M., Kyotalimye, M. and Tumbo, S. (2012) Tanzania. In: Waithaka, M., Nelson, G.C., Thomas, T.S. and Kyotalimye, M. (eds) *East African Agriculture and Climate Change. A Comprehensive Analysis*. IFPRI/CGIAR, Washington, DC, pp. 313–345.

Le Clerc, P. (2015). Adaptation des agriculteurs d'une vallée inondable à l'aléa climatique, Diagnostic agro-économique de la vallée du Kilombero, Tanzanie. Mémoire de master Recherche «pays émergents et en développement», UFR Agriculture Comparée et Développement Agricole, AgroParisTech/AFD, Paris, 147 pp.

National Irrigation Master Plan (NIMP) (2002) United Republic of Tanzania, (Ministry of Water and Irrigation, National Irrigation Commission), Dar Es Salaam.

Paul, J.-L., Duvail, S. and Hamerlynck, O. (2012) Appropriation des ressources «naturelles» et criminalisation des communautés paysannes: le cas du Rufiji, Tanzanie. *Civilisations* 60(1), 143–175.

Richard, P. (1993) Cultivation: knowledge or performance. In: Hobert, M. (ed.) *An Anthropological Critique of Development: The Growth of Ignorance*. Routledge, London, pp. 61–78.

SAGCOT (2012) *Investment Partnership Program, Opportunities for Investors in the Rice Sector*. Slideshow Presentation, October 2012.

Sandberg, A. (2004) Institutional challenges to the robustness of floodplain agricultural systems. In: *Paper Presented at the Third Penannual Workshop on the Workshop Conference,* Indiana University, 2–6 June 2004.

Stuhlmann, F. (1909) *Beiträge zur Kulturgeschichte von Ostafrika*. Deutsch-Ostafrika, Bd. X, Dietrich Reimer (Ernst Vohsen), Berlin, Germany, 905 pp.

Tewa, C. (2014) Favoriser la gestion du risque climatique par les agriculteurs et les pêcheurs: Diagnostic agro-économique de la Région de Kipo (zone de plaine inondable le long de le fleuve Rufiji–Tanzanie). Mémoire de fin d'étude, UFR Agriculture Comparée et Développement Agricole, AgroParisTech/AFD/ Sokoine University of Agriculture, Paris, 105 pp.

4 Managing the Flooding Hazard Through Hydraulic Planning: The Tonle Sap (Cambodia) and Mekong Delta (Vietnam)

Olivier Ducourtieux[1]*, Elsa Champeaux[2], Charlotte Verger-Lécuyer[3], and Florie-Anne Wiel[4]

[1]*Comparative Agriculture Training & Research Unit/UMR Prodig, AgroParisTech, Palaiseau, France;* [2]*Ginger SOFRECO, Clichy, France;* [3]*Cerfrance49, Saumur, France;* [4]*FADEAR, Bagnolet, France*

4.1 The Mekong River Basin: The Challenges of Global Climate Change

4.1.1 Geography of the Mekong: environmental, social, and economic issues

The 10th-longest river in the world (4900 km), the Mekong drains an 810,000 km^2 watershed inhabited by 70 million people in six countries (China, Burma, Laos, Thailand, Cambodia, and Vietnam). Its flow is governed by alternating monsoons, with a summer rainy season and a winter dry season. The river has its source in Tibet, in the Himalayas. However, snowmelt accounts for less than 10% of its final flow in the Mekong Delta[1] (MRC, 2005). Most of the water volume comes from precipitation on its course's lower half in tropical Southeast Asia, particularly from Laos onward, where the river's tributaries make up 70% of the flow at its mouth (Eastham *et al.*, 2008; Hoanh *et al.*, 2010a; Johnston *et al.*, 2010; Icem, 2013a).

The Mekong crystallizes the contrasts and contradictions of Southeast Asia: very large population (600 million inhabitants), sustained growth and emergence of new export-oriented economies (Thailand and Vietnam), critical biodiversity zone threatened[2] by deforestation, urbanization, and controversial hydroelectric planning on the river's course or tributaries (Molle *et al.*, 2009; Dugan *et al.*, 2010; Matthews and Gehab, 2014).

4.1.2 Climate change in the Mekong Basin

Studies on the impact of global warming on the Mekong watershed, and more generally on the Greater Mekong region, are recent or ongoing, with partly divergent projections (Eastham *et al.*, 2008; Hoanh, 2010a,b; Johnston *et al.*, 2010; Lacombe *et al.*, 2010, 2013; Icem, 2013a,b, 2014a,b,c,d; IPCC, 2014b). However, the IPCC[3] aligns with the authors' consensus toward increased precipitation over Southeast Asia (IPCC, 2014b, p. 1355). Surface runoff in the Mekong watershed (Fig. 4.1) could rise by a factor of 2.5 in the height of the rainy season, with increased interannual variability (Eastham *et al.*, 2008; Lacombe *et al.*, 2010; Icem, 2013a; Teng *et al.*, 2016).

Compared to other parts of the world, climate change in the region will then result in a limited

---

*Corresponding author: olivier.ducourtieux@agroparistech.fr

©2024 CAB International. *Agrarian Systems and Climate Change: Journeys of adaptation in the Global South* (eds H. Cochet, O. Ducourtieux, and N. Garambois)
DOI: 10.1079/9781800628137.0004

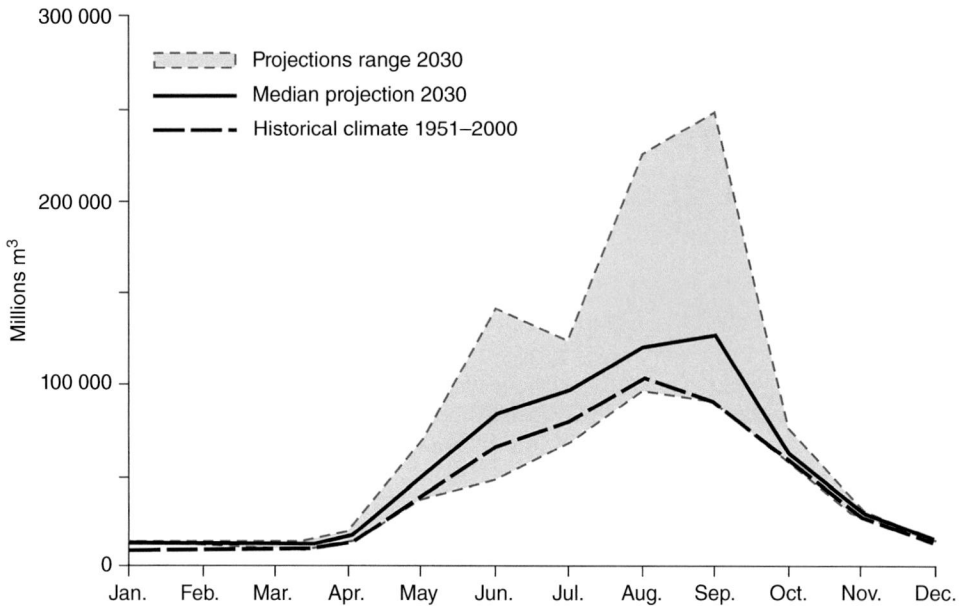

Fig. 4.1. Runoff in the Mekong watershed (Eastham *et al.*, 2008, p. 8, © Copyright CSIRO Australia).

rise in temperature of 1.5–2.5°C by 2050 (Icem, 2013b; IPCC, 2014a,b), increased precipitation, Mekong River floods in intensity, with a higher frequency of extreme events, especially floods. These changes are bound to impact human activities, be it agriculture (Icem, 2014a,c), fishing (Icem, 2014b), or others (Icem, 2014d).

In Cambodia and Vietnam, the general trends described for the watershed remain applicable at the local level (Arias *et al.*, 2012; Roth and Grunbuhel, 2012; Murphy *et al.*, 2013; Doch *et al.*, 2015), with effects already observable (Ham and Someth, 2015):

- An increase in rainfall, especially at the height of the rainy season.
- A change in annual precipitation distribution, going from a bimodal regime with a relative decrease in rainfall in August to a monomodal regime with a peak in September.

In the Mekong Delta, modeling converges on similar scenarios, with a moderate increase in precipitation of 0.4% by 2030 and 1.2% by 2080. Annual mean temperature would rise by 0.6°C in 2030 and 1.8°C in 2080 (Teng *et al.*, 2016). The sea level is currently rising about 3 mm/year. In the future, the trend could accelerate with projections of +1 m in 2080 due to the thermal expansion of the China Sea (MoNRE, 2009; Smajgl *et al.*, 2015).

Through the combination of their exposure to climate hazards (namely, the increased risk of flooding) and low economic capacities for adaptation, Cambodia and Vietnam are ranked among the most climate change-prone countries by many authors (Kreft *et al.*, 2015).

For farmers in our study areas, climate hazards will not change in nature but will increase in frequency and intensity. Their economies' resilience will be affected, with reduced recovery intervals between two extreme events, such as catastrophic floods.

4.2 Diversity, the Keystone of Resilience and Adaptation in a Human-Built Environment with Uncontrolled Hazards

4.2.1 The Tonle Sap floodplain in Cambodia

The study area[4] is located southwest of the city of Kampong Thom, the capital of the eponymous province in Cambodia (Fig. 4.2), on the shores of Tonle Sap Lake. This lake's hydrology is unique (Kummu *et al.*, 2014):

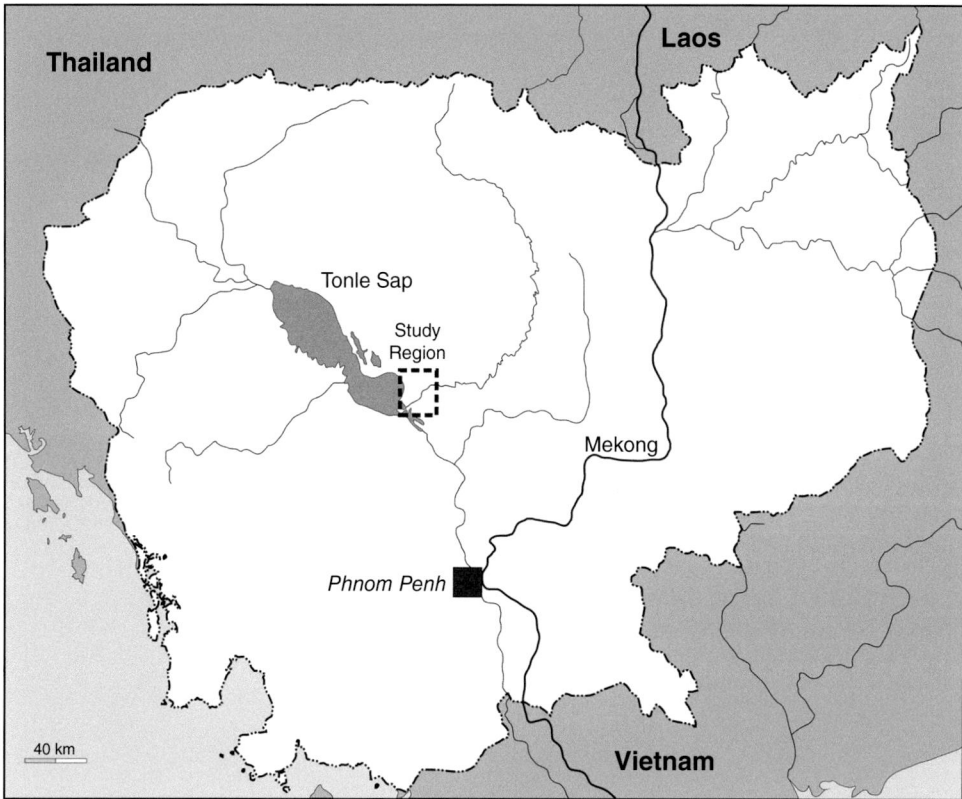

Fig. 4.2. Mekong River and Tonle Sap Lake in Cambodia.

- In the dry season, the lake covers 2700 km² with an average depth of 1.5 m. It feeds the Tonle Sap River, which flows into the Mekong River in Phnom Penh.
- In the rainy season, the Mekong's flow increases tenfold, and its level becomes such that the water circulation reverses in the Tonle Sap River, filling the lake, which becomes an outlet. The lake's area is multiplied by six, reaching 16,000 km²; that is, 10% of Cambodia's surface area. Its flooding merges with that of the Mekong to cover more than half of Cambodia's central plain.

In the lake, the average annual flow reaches 45 Gm³ of water, 54% of which comes from the Mekong, 29% from the lake's tributaries, and 14% from direct rainfall (Matsui *et al.*, 2006). Due to the extent of the Tonle Sap flood, the area covered by potentially highly fertile recent alluvium is vast (32,000 km²).

This region has a long history of cultivation, with the emergence of a powerful centralized State as early as the 9th century, whose power was based on the wealth of irrigated and flooded rice cultivation: the Khmer hydraulic empire of Angkor (9th–13th century). Among the hypotheses explaining this empire's decline and that of a multi-year drought highlights the social, economic, and political importance of climatic hazards in the region (Nuorteva *et al.*, 2010).

In addition to strong seasonal variation in the Mekong River water regime, there is also a very marked interannual variability related to the basin's rainfall variability (Eastham *et al.*, 2008; Icem, 2013a; Fig. 4.3).

Local rainfall, averaging 1500 mm per year, is concentrated from May to October (85%, Fig. 4.4), with high variability at the beginning (April–May) and end (November) of the rainy season. This variability is a risk factor for crops

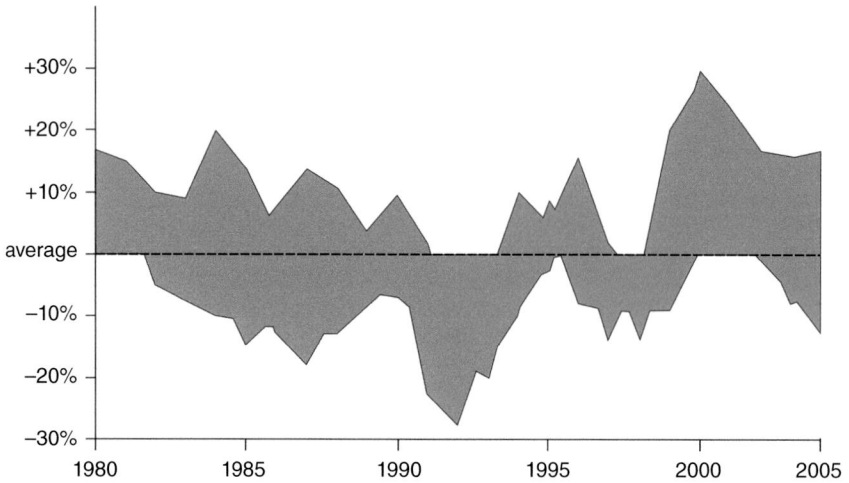

Fig. 4.3. Precipitation variability in the Mekong River Basin. Deviation of annual totals from the 1980–2005 average (Icem, 2013a, p. 41).

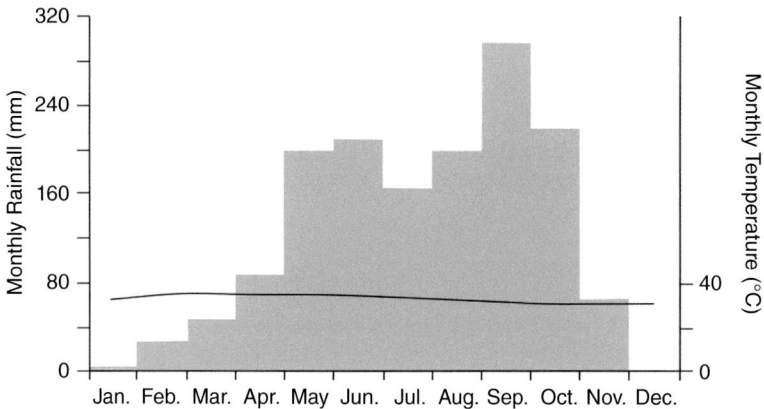

Fig. 4.4. Ombrothermic diagram of Kampong Thom station in 2000–2010 (Lécuyer and Wiel, 2014, p. 10).

on fields in a "high" topographic position, not submerged by the river flood.

Farmers in the region take advantage of the Tonle Sap's high water level and the time lag between local rainfall and flooding to develop a complex rice-growing calendar, with two rainy-season crop cycles, a flood recession cycle, and two dry-season irrigated cycles (Fig. 4.5).

However, they have to deal with the uncertainty of the timing and level of the water rise, as well as the risks of flooding crops or drought depending on the year, to provide for their household.

4.2.2 Small-scale environmental heterogeneity in a highly human-built area

The region studied is characterized by an absence of relief: less than 15 m of difference in altitude between the lake's bottom and the highest points on land (e.g. the banks of the national road). However, because of the extent of the Tonle Sap's annual flooding, micro-reliefs are of considerable importance to farmers for making use of the ecosystem.

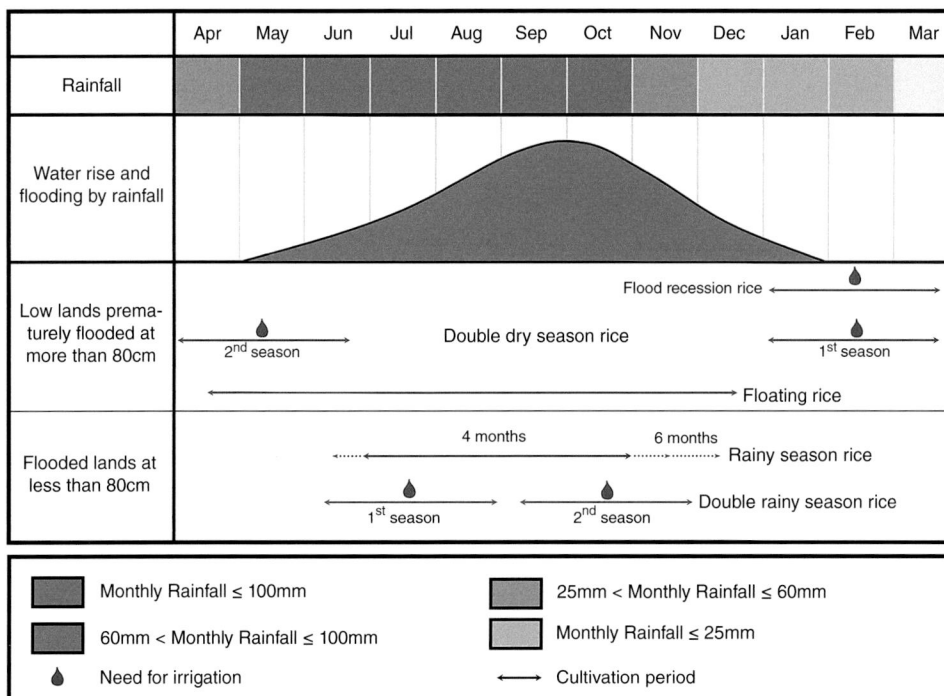

Fig. 4.5. Precipitation, flood regime, and crop cycles (Lécuyer and Wiel, 2014, p. 18).

While the uninitiated visitor struggles to discern differences in this vast plain during the dry season, five different agroecological stories are visible (Lécuyer and Wiel, 2014) based on their relative elevation (Figs. 4.6 and 4.7):

- The lake's minor bed, always underwater even in the height of the dry season. It is a space dedicated to fishing, as well as housing the fishermen's floating villages (Photo 4.1).
- The alluvial forest emerged in the dry season and flooded under 3–10 m of water in the rainy season. This space is devoted to fishing and timber harvesting, and the forest is tending to regress due to the extension of areas cultivated with floating rice since 1979.
- The low plains, vast grasslands grazed in the dry season and flooded in the rainy season. If the water level does not exceed 3 m, it is possible to cultivate floating rice.[5] As they are the last dry lands, it is also possible to grow rice during the recession flood, irrigated from water retention basins dug a little higher.

- The high plains, which are unflooded in the dry season, are generally slightly and temporarily flooded in the rainy season. Grazed grasslands in the dry season, they are planned in rice paddy fields in irrigated schemes, with one or two annual crop cycles depending on the floodwater's height. The highest areas (12–15 m above sea level) have always emerged, except in the case of catastrophic flooding, and are used for transport (roads and paths), housing settlements grouped into villages, or a few orchards (especially mango trees).
- The Stung Sen levee,[6] still emerged, where a road passes and villages are established. From there, farmers can fish in the river or pump water to irrigate the rice fields below, on the back side of the riverbank.

Originally forested, the region has long been strongly transformed by man, with a high level of anthropization, culminating in the planning of irrigated schemes.

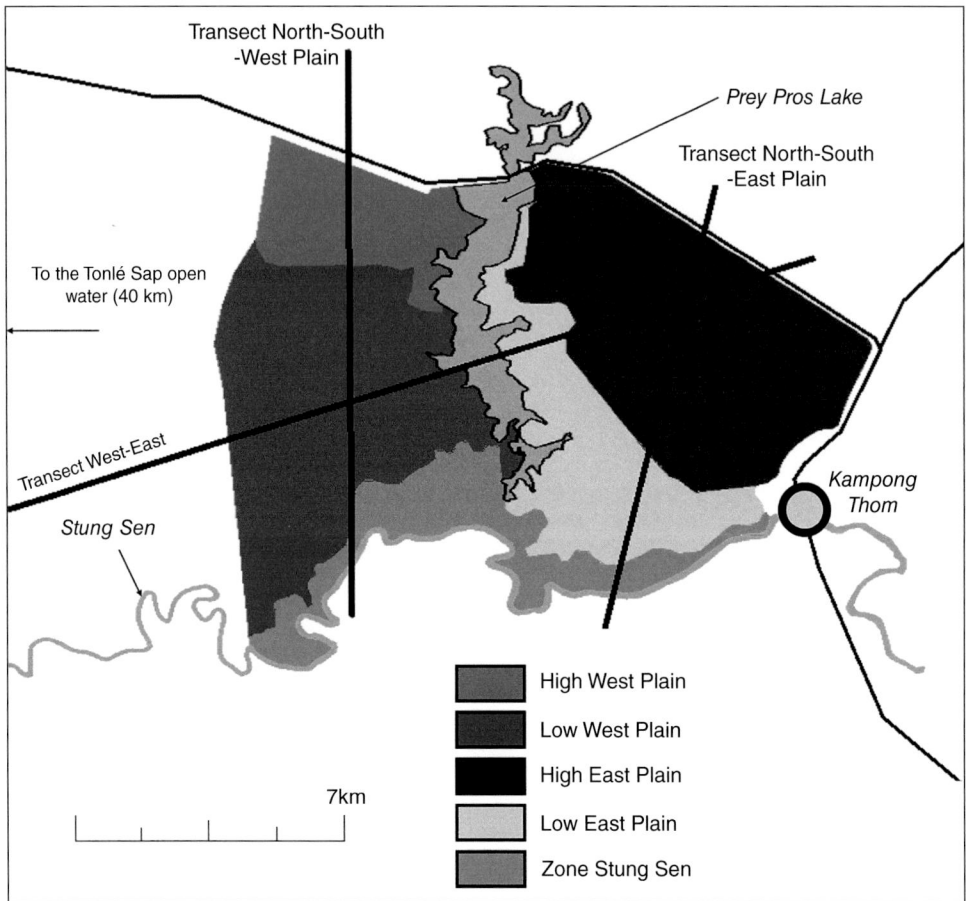

Fig. 4.6. Zoning of the western region of Kampong Thom (Lécuyer and Wiel, 2014, p. 17).

4.2.3   Resilience of small-scale agriculture: a multi-scale adaptation to hazards

4.2.3.1   Risk management at the field level: rice cultivation

While corn monoculture is a major risk factor in the face of climatic hazards in the Zambian land (Chapter 2, this volume), the combination of crop species on the same field is a prevention and mitigation strategy in the Rufiji River Valley in Tanzania (Chapter 3, this volume).

Rice monoculture has long been the standard farming practice for developing the Tonle Sap floodplain. The tolerance of *Oryza sativa* to partial or temporary submersion allows for the cultivation of inundated areas while eliminating the task of weeding and some sanitary risks (especially nematodes). Secular mass selection in the region has made it possible to reinforce this tolerance and obtain a sampling of varieties with characteristics finely adapted to the ecosystem's various stories. Farmers in the study area thus have three floating rice varieties (photoperiodic), several dozen rainy-season rice varieties (photoperiodic, two dominant cultivars), and two dry-season rice varieties[7] (non-photoperiodic).

In the rainy season, floating rice varieties allow farmers to cultivate fields submerged under a few meters of water (1–4 m). In the same season, farmers grow photoperiodic varieties with different ripening dates to utilize the environment's different stories according to the likely dates of flood arrival and retreat and altitude. In addition, on some lands that are limited in terms

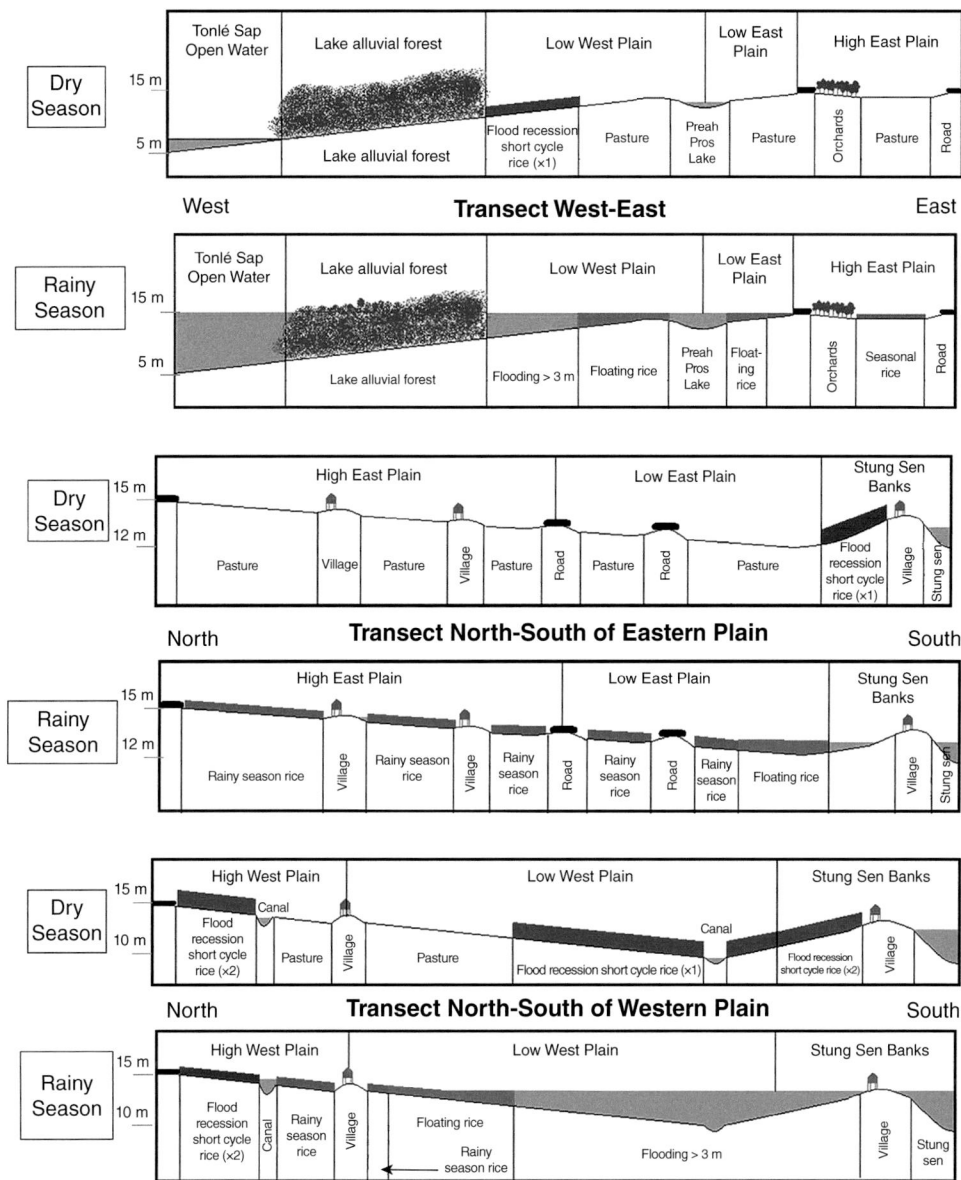

Fig. 4.7. Toposequence of agroecological stories of the Tonle Sap riverbanks (Lécuyer and Wiel, 2014, pp. 16–20).

of water level in the rainy season, farmers plant several varieties of rice in order to limit risks. This also has the effect of staggering the sowing, transplanting, and harvesting periods (Lécuyer and Wiel, 2014). Non-photoperiodic and high-yielding varieties are reserved for fields with maximum water control: dry season or recession flood irrigated rice fields.

Transplanting, practiced in particular for rainy-season rice cultivation (excluding floating rice), is also a risk-limitation strategy. It allows the flooding risk to be anticipated. From May to August, rainfall is sometimes insufficient to satisfy a rice crop. Transplanting provides protection against drought risk, since the rice is then in a single field suitable for gravity irrigation. In addition, transplanting allows great adaptability to climatic hazards. Thus, several levers are used by farmers to adapt to the changing climate (Lécuyer and Wiel, 2014):

- Sowing date in the nursery. As tilling is to be done on a smaller surface, it is easier to do it quickly, and it can be done from the first rains.
- The transplanting date can be adapted to the height of the water in the field to be transplanted but also to the degree of maturity of the rice to be transplanted or to the height of the water in the nursery.

Transplanting is thus an anti-random strategy that maximizes the harvested area in relation to the cultivable area, at the cost of increased labor. In addition to the differences in yields, the resilience of such practices to climatic hazards is exemplary. This explains why it is still very present in the landscape, despite the significant lack of labor force. Farmers keep as much transplanted area as labor force availability allows. However, irregular rainfall at the beginning of the wet season sometimes results in dry or flooded fields. In such cases, planting must be repeated. If the flooding occurs late (shortly before transplanting), it is no longer possible to prepare other seedlings in the nursery. Transplanting is then completed by direct seeding on the highest, least flooded land (Lécuyer and Wiel, 2014).

Rice cultivation on the banks of the Tonle Sap is therefore not a normative monoculture. It is a great diversity of cropping systems (Table 4.1) with differentiated performances[8] carefully designed and practiced by farmers to astutely adapt to the region's microrelief, random water regime, and available production factors: labor force (staggered work), land (two cycles on the same field, cultivation of all the agroecological stories), and capital (for irrigation).

Among the rice-growing systems, SRI (System of Rice Intensification) is one of the most recent (Lécuyer and Wiel, 2014). However, it is not widely used in the region; the principle of transplanting very young (5–10 days), and therefore short seedlings, increases the risk of loss in the event of rapid and early flooding. The partial control of water in the rainy season, which is likely to weaken with the strengthening of climatic hazards, limits the possibility of developing SRI in the floodplain during the rainy season. While this cropping system can increase yields at a limited cost, it is only possible with increased labor from a labor force which is not always available, and increased sensitivity to climatic hazards affecting water control.

4.2.3.1.1 RISK MANAGEMENT ON THE PRODUCTION SYSTEM LEVEL: ACCESS TO DIFFERENT ECOLOGICAL STORIES AND COMBINATION OF CROPPING SYSTEMS. Like farmers in the Rufiji Valley in Tanzania

Table 4.1. Different rice production systems of the Tonle Sap riverbanks in Kampong Thom (Lécuyer and Wiel, 2014, p. 95).

Cropping system	Cycle (months)	Zone	Field size (ha)	Yield (t/ha)
Floating rice	6 (5–12)	Low East Plain	<0.5	1–1.5
Floating rice	7–8 (5–1)	Low West Plain	>1	1.5–2
Rainy-season rice	4 (7–10)	East-West High Plains	0.3–1	2–2.5 (with transplanting)
Rainy-season rice	6 (7–12)	East-West High Plains	0.3–1	1.2–1.5 (direct drilling)
Double rainy-season cycle, cycle 1	3 (6–8)	High West Plain (irrigated)	1 (subdivided into 2–4)	3.5
Double rainy-season cycle, cycle 2	3 (9–11)	High West Plain (irrigated)	1 (subdivided into 2–4)	3
Dry-season rice	3 (1–3)	Low West Plain	<0.5	7
Double dry-season cycle, cycle 1	3 (1–3)	Stung Sen (irrigated)	<0.5	4–5
Double dry-season cycle, cycle 2	3 (4–6)	Stung Sen (irrigated)	<0.5	2–2.5

(Chapter 3, this volume), those on the banks of the Tonle Sap River seek to combine several rice-growing systems in different agroecological stories in order to:

- spread out workload peaks. This makes it possible to cultivate a larger area or, when this is limited, devote the family labor force to other activities (animal farming, other crops, fishing and hunting, agricultural wage-earning jobs, migration to distant wage-earning jobs); and
- limit risks in the face of climatic hazards, whether destructive flooding or drying up fields, depending on the timing of the flood's arrival and retreat and its intensity.

However, most households do not have access to every ecosystem. In the region, socioeconomic differentiation has historically been based on this ability to cultivate several stories.

In the villages that existed before the Khmer Rouge took power in April 1975, social and economic differences were partially erased by this genocidal regime. Indeed, 2.2 million people[9] (out of 7.9 million inhabitants) died in 3 years from hunger or exhaustion from grueling work in poorly designed rural cooperatives and without taking into account the opportunity cost of farmer labor.[10] Others were executed (Heuveline, 2001; Pillot, 2007). The population was displaced on a large scale in the country.

It was not until 1979, after the overthrow of Pol Pot by the Vietnamese, that the first families among the survivors returned to the Kampong Thom region, while those who had occupied it went back to their original villages. These first "returnees" took over the few surviving buffalos. The Vietnamese occupier and the new State quickly established a collective organization less coercive than the Khmer Rouge "cooperatives," the Krom Samaki.[11] During the rainy season, rice fields were cultivated collectively. However, households with a large number of laborers or those that recovered a buffalo could also divert a fraction of their labor force to clear the alluvial forest. They then cultivated their own floating rice fields, allowing them to begin a process of early accumulation of land and rice. These savings were then quickly reinvested in buffalos, then power tillers and pumps, to increase labor productivity. These families cleared more and more of the alluvial forest to increase the area

under cultivation, thus rapidly widening the income and food security gap with families who did not initially have this supernumerary labor force.

Although farms tend to become fragmented when inherited,[12] the largest farms with the highest economic results today are those of "first-time farmers" or their descendants. Among the dozen or so production systems identified in the region, the difference in surface area is significant (from 0.2–3 ha), as is the difference in farm income: from 30 to 1000 US$ total per laborer (300–1600 US$ per laborer for total annual income).

For migrants recently settled in the area, access to land is difficult, except for families with capital. These households settle on the levee of the Stung Sen and invest in rice fields that can be irrigated in the dry season, thus largely freeing themselves from climatic risks. Without capital, they have to rent expensive rice fields or settle on the most marginal lands (low plain) with very small fields. The risk linked to climatic hazards is maximum there.

4.2.3.1.2 MANAGING RISK AT THE HOUSEHOLD LEVEL: OFF-FARM ACTIVITIES, INCLUDING MIGRATION, TO REDUCE POVERTY AND MITIGATE RISK. Agricultural income represents at most three-quarters of total household income, and sometimes less than one-fifth. To supplement this income, family labor is mobilized for activities other than rice cultivation, such as:

- vegetable gardens around the house (all households) or, on a larger scale, market gardening on unflooded land (high plain and bank of the Stung Seng);
- cultivation of fruit trees (mango and cashew) around the house (all households) or orchards of a few acres for families with access to emerged land in the high plain;
- poultry (all households), pigs (very limited), and cattle (buffalos and cattle, both for tillage and for the meat market) in limited numbers (1–3 heads/household);
- fishing (in ponds, canals, and the Stung Sen) and hunting (capture of rodents and insects);
- day laborer job for the poorest families[13] and provision of services (renting out draft animal hitching equipment or power tillers, combine harvesters) for the wealthiest; and

- service activities (transportation) and commercial activities in the village (grocery store) or in Kampong Thom for families with sufficient cash flow.

In varying proportions, all of these activities allow farmers' families to supplement their work calendar and income. Few of them would be able to make a living from rice cultivation alone, whereas the accumulation of different activities confers a total income per laborer that is higher than the opportunity cost of labor in the region (Lécuyer and Wiel, 2014).

In addition to contributing to income, diversification of activities reduces vulnerability to climate risks. Combining rice-growing systems in different agroecological stories and seasons with other activities meets the dual objective of increasing labor productivity and reducing vulnerability to flooding or drought.

This logic is taken a step further in many families, with some of the youngest workers migrating outside the study area to access jobs. While the wealthiest families use their networks to access remunerative opportunities (trade) or opportunities that bring power (public service), the poorest are limited to the least remunerative salaried jobs:

- In Phnom Penh, industry (clothing, construction, etc.) and services (housemaids and guards, prostitution, etc.).
- In the hilly regions of the west, northwest, or northeast, on the agricultural frontier, as day laborers on concessions (logging, rubber plantations, or the most well-off farms (because they arrived early) (Diépart and Sem, 2014).
- In Thailand as employees (rubber plantations, sea fishing), in basic industrial or service jobs. Often of precarious status, this expatriate workforce is dependent on the vagaries of the economic and political situation in Thailand.

For the poorest families, these migration-related jobs contribute only marginally to household income. Often, the young people who went to the capital returned to the village regularly and left with rice given by their parents to survive in the city. On the other hand, these jobs make it possible to reduce the impact of land fragmentation during inheritance in order to maintain sufficient land for the household, doing so in different agroecological stories. They can also serve as backup in case of extreme events leading to the abandonment of the family farm.

4.2.3.1.3 THE DESEASONALIZATION OF RICE CULTIVATION THROUGH DRY-SEASON IRRIGATION: A STRATEGY FOR REDUCING VULNERABILITY CONDITIONED BY ACCESS TO CAPITAL. Since the early 2000s, families have gradually invested in moto-mechanization based on multi-purpose power tillers, used for tillage (Photo 4.2), transport (Fig. 4.8), or irrigation (Fig. 4.9).

For families with the means to acquire them, these machines make it possible to increase labor productivity. They also make it possible to reduce

Fig. 4.8. Power tiller for tillage and transport (Ducourtieux, 2014).

Fig. 4.9. Power tiller-driven irrigation pump (Ducourtieux, 2014).

risks associated with planting rice at the beginning of the rainy season (from uncertain flood timing and intensity) through faster and more flexible (dry or wet plowing) motorized labor. For rainy-season rice, farmers often have to wait for the first rains to plow, without delaying sowing too much. Otherwise, the risk of flooding the crop at the beginning of the rainy season is increased (Lécuyer and Wiel, 2014).

Although it improves the fertility of the local soils, the Tonle Sap's water rise is a source of uncertainty and risk for farmers because of its unpredictable timing and intensity. This hazard is expected to intensify. To reduce their vulnerability, some families have radically changed their cropping calendar for all or part of their rice fields, which are now cultivated in the dry season (two cycles, Table 4.1). This involves the use of irrigation.

This approach is only possible for farmers with easy access to water: proximity to canals in the west low plain, back of the Stung Sen riverbank (Fig. 4.7 and Table 4.1). They must also have the means to invest in a pump and power tiller (Fig. 4.9). Disconnected from the flood regime, this cropping system is very safe; it is also the most value-added, both per unit area and labor.

Expansion of the irrigation network could increase access to dry-season rice cultivation. This is the motivation for public investment in the irrigated areas in the northwest of the study area and on the left bank of the Stung Sen River in the south of the study area (Figs. 4.10 and 4.11). However, these public infrastructures seem to be encountering prohibitive difficulties: while rehabilitation works have come one after another since their creation during the Khmer Rouge era, they have been totally or partially out of service for years. The scale of the infrastructure leads to complex management and impossible operating and amortization costs for farmers.

Masonry pipe from the Khmer Rouge era, primary canal recently rehabilitated.

Further west and south of the study area, large-scale planning can be observed (Fig. 4.12). These are concessions granted by the State to private investors to build retention basins filled by the flood, irrigating rice-growing areas of several hundred hectares during flood recession (Fig. 4.13).

Fig. 4.10. Channels for collective irrigation (Ducourtieux, 2014).

The study area is located inside the box; the arrows highlight the rice concessions, visible on the satellite image (source: Google Earth).

Initiated in 2008, the project was designed to give one million hectares to the Emirate of Kuwait in exchange for oil (a $546 million contract). Faced with growing criticism of international land grabs, the Emirate withdrew. The Khmer government then allocated these areas[14] to national investors, mostly from Phnom Penh, who are well connected to the government's clientelist networks. These investors are attracted by the high prices of rice on the world market.[15] Although they produce less value-added per area than farmer activities,[16] these projects exclude villagers from productive spaces that can be planned to improve their standard of living and reduce their vulnerability to climate risks.

Fig. 4.11. Set of pumps for the irrigation of a collective irrigated scheme from the Stung Sen (Ducourtieux, 2014).

4.2.3.1.4 CONCLUSION. On the banks of the Tonle Sap in Cambodia, farmers have long developed complex rice cultivation in a highly artificial environment, with a wide range of varieties for:

- farming the largest possible area of the floodplain and its stories differentiated by the microrelief. They grow floating, flood recession, and rainy-season rice in one or two cycles, and double-cycle irrigated rice in the dry season;
- limiting the risks of climatic hazards, including flooding's calendar uncertainties, flood intensity, and drought; and
- maximizing labor productivity.

From the beginning of the 1980s, households that had better weathered the vicissitudes of the country's recent history (civil war and the Khmer Rouge regime in particular) were able to accumulate capital. The capital was quickly mobilized to cultivate the largest possible areas in complementary agroecological stories and invested in moto-mechanized farming equipment (power tillers and pumps). The equipment contributes to securing agricultural production in the face of climatic hazards by conferring greater flexibility on the work calendar due to increased speed. It has also allowed rice cultivation to be deseasonalized (dry-season irrigation).

Dry-season irrigation appears to be a preferred approach to reducing vulnerability to a climate hazard that is expected to intensify. However, the concentration of irrigable land in the hands of the most affluent households, and especially in those of urban investors from outside the region, is to the detriment of the most vulnerable families who only have access to the areas most exposed to flooding. For the latter, some of the family's laborers emigrate to Phnom Penh and their secondary and tertiary sectors, some to the agricultural pioneer fronts of the hills in the west and north of the country or to Thailand. This contributes to increasing the total household income and diversifying its sources, thus reducing their vulnerability.

4.3 Extensive Environmental Artificialization: A Hazard Under Control for Strong Development Dynamics, with An Uncertain Future

4.3.1 Between land and water, the Mekong Delta

At 55,000 km², the Mekong Delta is one of Asia's mega-deltas (Fig. 4.14) with 18 million

Fig. 4.12. Large rice field of private investors (Ducourtieux, 2014).

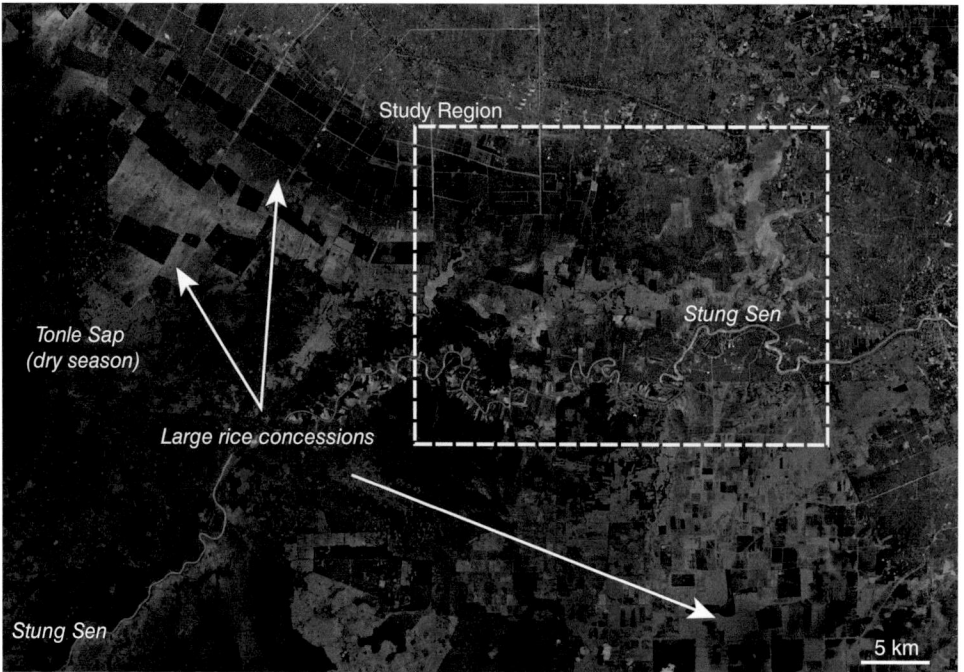

Fig. 4.13. Satellite image of the Kampong Thom region.

people living there (330 people/km², 20% of Vietnam's population), cultivating 7.5 million hectares of land formed by river alluvium accumulated over 6000 years (Takagi *et al.*, 2016).

Although it borders the megalopolis of Ho Chi Minh City,[17] the Mekong Delta contributes less than 10% to the country's industrial production. Yet the region is crucial to the country's economy because of its inhabitants' agricultural and aquacultural production.

Rice cultivation is a key component of the Vietnamese economy and culture. Despite history's vicissitudes, national production has been growing since the late 1970s and the country has been food sovereign since the early 1990s (Fig. 4.15). It is beginning to emerge as a major player in the global rice market, competing with Thailand as the second largest exporter, behind the USA. Today, Vietnam exports about 15% of its annual production and makes up 12–15% of world trade (Fig. 4.16).

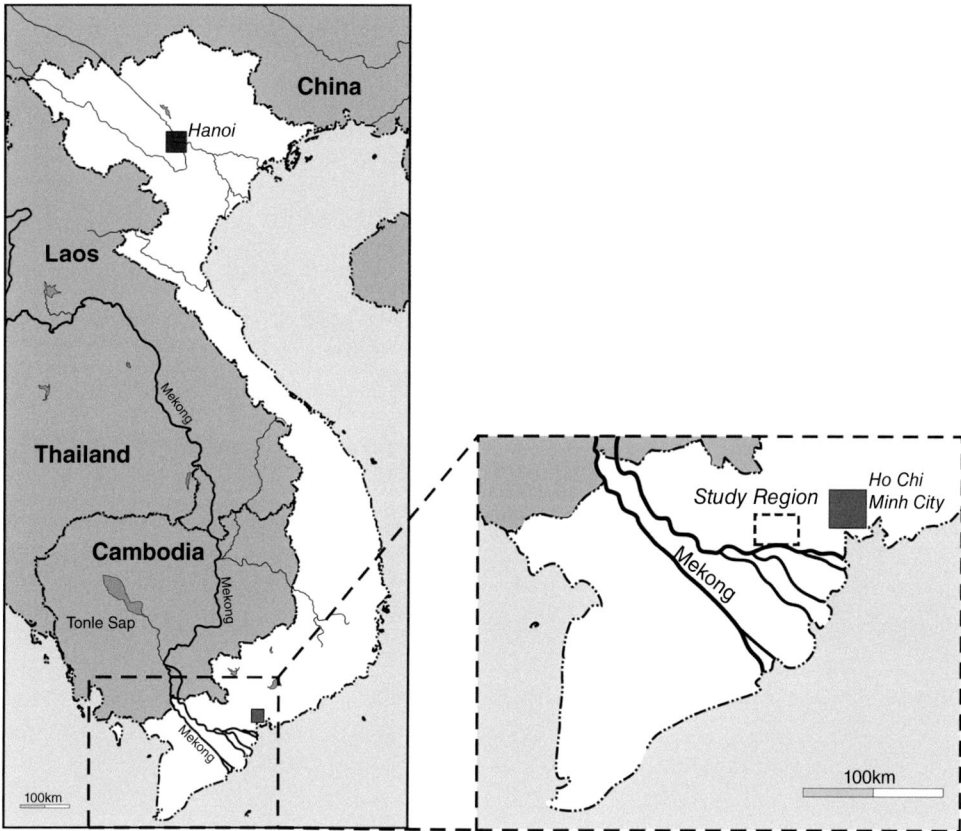

Fig. 4.14. The Mekong Delta in Vietnam.

Mekong Delta farmers are central to these results: with 25 million tons of paddy rice produced in 2015, the Delta accounts for 57% of national production (General Statistics Office, 2016). Accounting for less than 3% of the world's rice area, Mekong Delta production covers 10–12% of international trade, contributing to the food security of many grain-deficient regions of the world. A possible contraction of rice harvests in the Delta—due to climate change, for example—would have deleterious global effects that go far beyond the region's borders.

With the mouth of the Mekong River in the extreme south of Vietnam, the Delta is a floodplain of low altitude (0.5–5 m above sea level, average 0.82 m, Minderhoud *et al.*, 2019), crossed by the Mekong's four arms and a very dense network of tributaries and canals. The hydrology is complex and aggregates:

- significant (1400 mm/year) and seasonal rainfall (Fig. 4.17)
- the river's seasonal flow (2500–27,000 m³/s) depending on rainfall over the 810,000 km² watershed
- tides with a wave influence that starts decreasing 100 km from the river's mouth

The Mekong flood is ambivalent. Devastating when it exceeds the level of the levees isolating the land from the watercourse, it contributes to the Delta's alluviation and draining of the brackish water that reaches land in the dry season. Historically essential for rice field irrigation in the rainy season, it requires significant subsequent hydraulic planning and skillful water management to drain and enhance the lowest areas of the floodplain.

Thiện Trí commune in Cái Bè district, Tiền Giang province, was selected as the study area

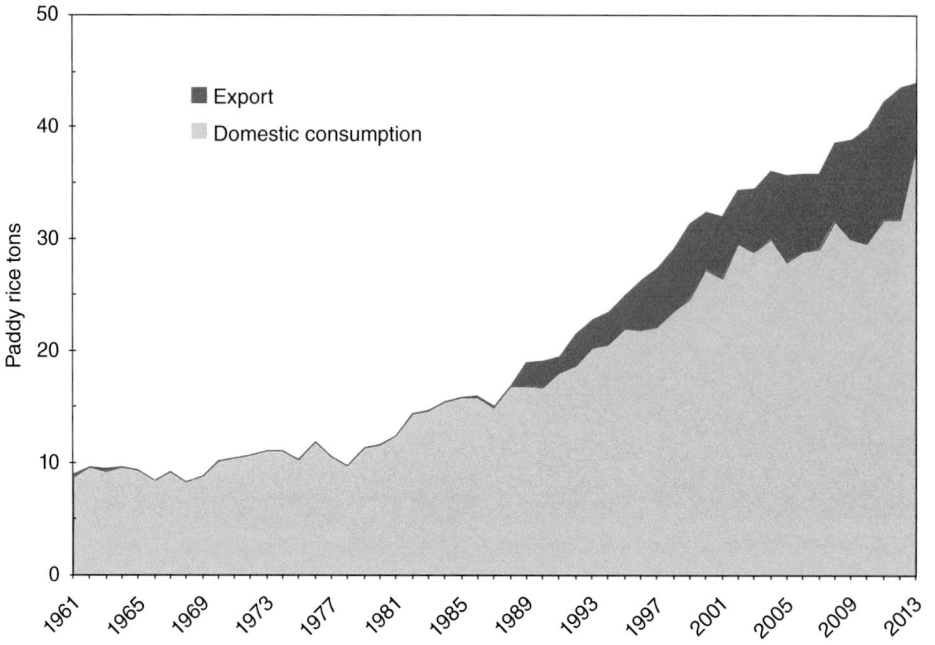

Fig. 4.15. Evolution of rice production in Vietnam (FAOStat, 4/2017).

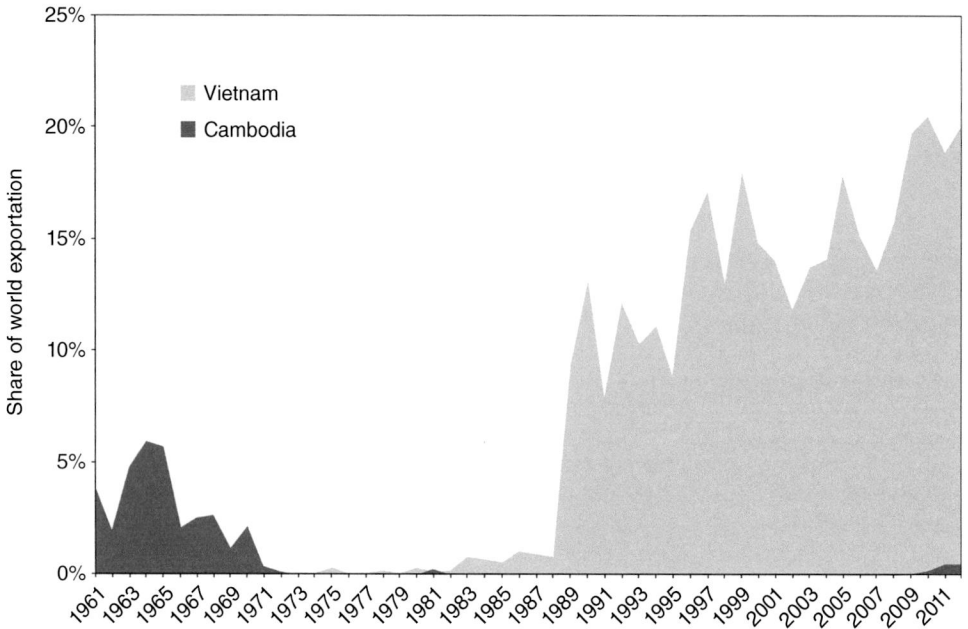

Fig. 4.16. Vietnamese export share in the global rice market (FAOStat, 6/2023).

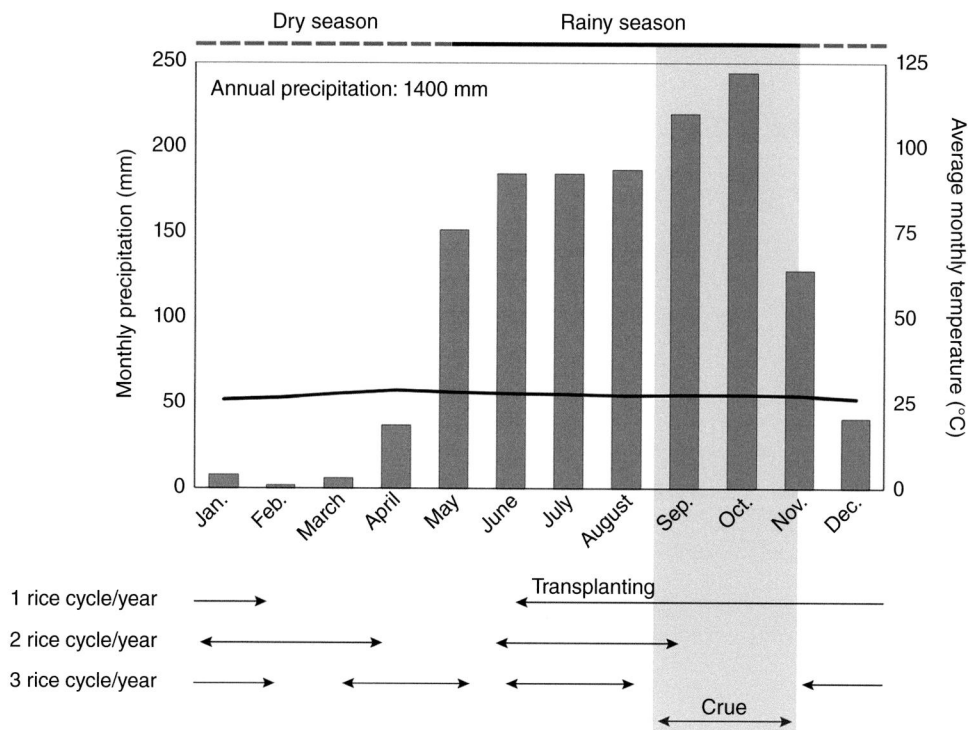

Fig. 4.17. Ombrothermic diagram of Cái Bè and rice cultivation calendars, average 1982–2012 (Champeaux, 2016; Climate-data.org).

(Figs. 4.8 and 4.18).[18] Due to the remoteness of the river's mouth, the surface water is fresh with no seawater intrusion. However, waves from the tide are very pronounced in the area, with a reversal of flows twice a day in the rivers. Amplitude exceeds 2 m between high spring tide and low tide.

In the floodplain, a very small slope (0.03% on average) distinguishes the sandier Mekong natural levee from the lower, more clayey areas, with a difference in level of 2 m. This microrelief, which is not very noticeable, explains the land's differentiated exposure to flooding during the water rise (September–November) and the daily tide cycle (Fig. 4.19):

- To the south, the top of the levee is still emerged. Dwellings are established there, as well as animal farming and orchards.
- To the north, the lowest lands, which are swampy, are flooded regardless of the tide. It is only possible to develop them if they are dammed up to isolate them from the flood.

- Between these extremes, the southernmost lands are flooded only during high spring tides.[19] A large central portion is flooded twice a day at high tide. The development of this land depends on the ability to use the tide's daily alternation to drain and irrigate fields, especially for rice cultivation.

Levees and dikes reproduce the same sequence between the main channels and the lowest areas of the plain.

4.3.2 The control of water for rice production, a political constant through the vagaries of history

At the end of World War II, Thiện Trí commune was mainly cultivated with rice. The annual cycle began with the rainy season for yields of about 2 tons/ha of paddy (Fig. 4.20). The top of

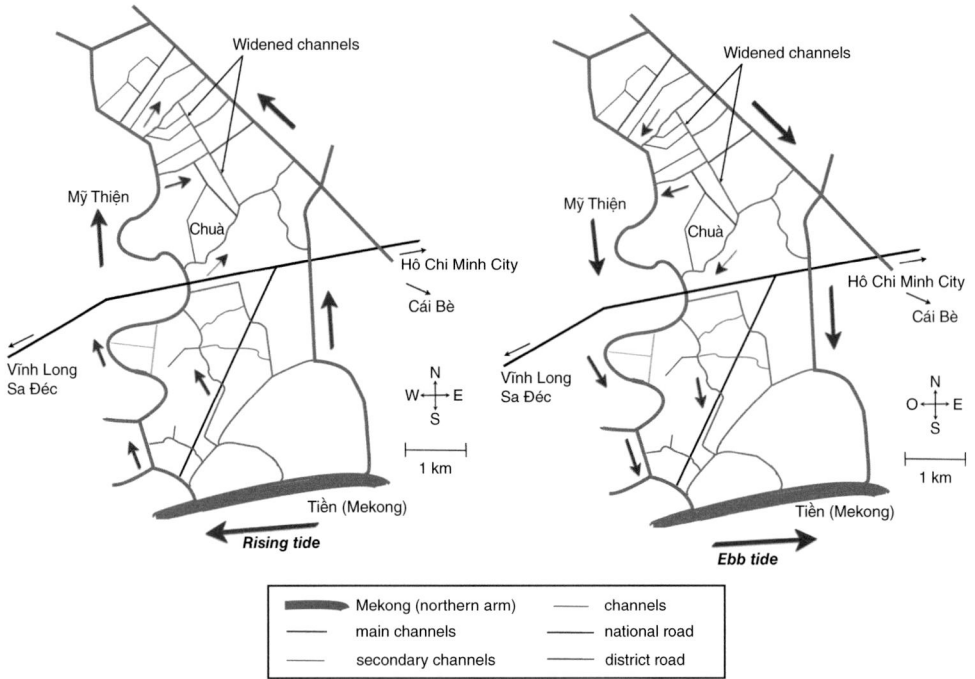

Fig. 4.18. Tides and flows in the channels of Thiện Trí commune, Mekong Delta (Champeaux, 2016, p. 12).

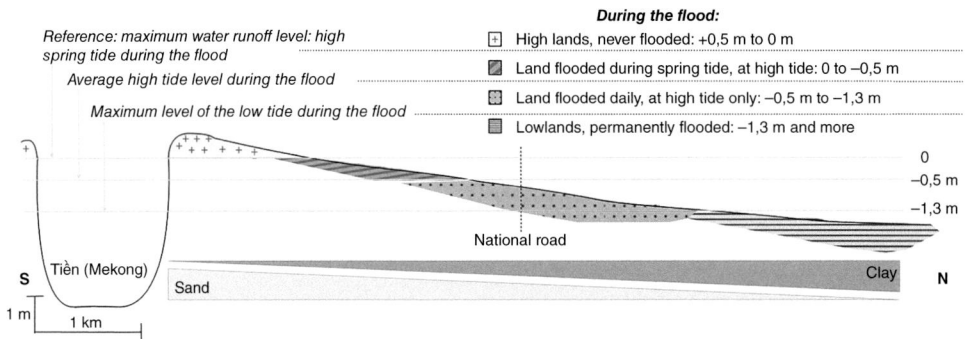

Fig. 4.19. South-north cross-section of Thiện Trí commune and water levels during the flood (Champeaux, 2016, p. 22).

the natural levee was forested. Farmers supplemented their income and food with gardens and raised small farm animals around their houses, always on the levee.

Water management was basic, with a few wooden gates through the channel levees. When water levels were not extreme, these gates allowed the paddy fields to be irrigated at high tide and drained at low tide (Fig. 4.21). In the dry season, the water level was too low, even at high tide; the

rice fields were not cultivated, and the animals grazed this space freely. When the water level was significant, it rose above the levee and covered the land to the detriment of yields if the flooding continued. Flooded during the 2–3 months of rising water, the lowest areas to the north were swampy and not cultivated (Champeaux, 2016).

The population's social differentiation was very marked in the region. More than two-thirds of the population rented small areas (0.2–3 ha/

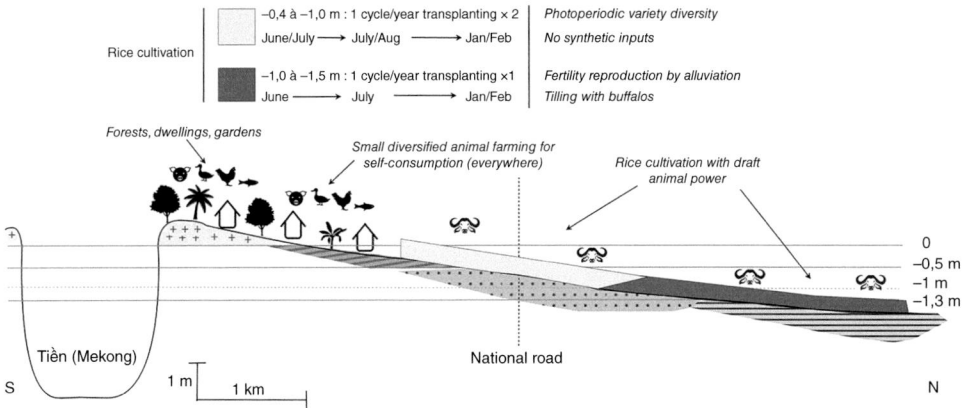

Fig. 4.20. South-north toposequence of Thiện Trí commune in the late 1950s (Champeaux, 2016, p. 30).

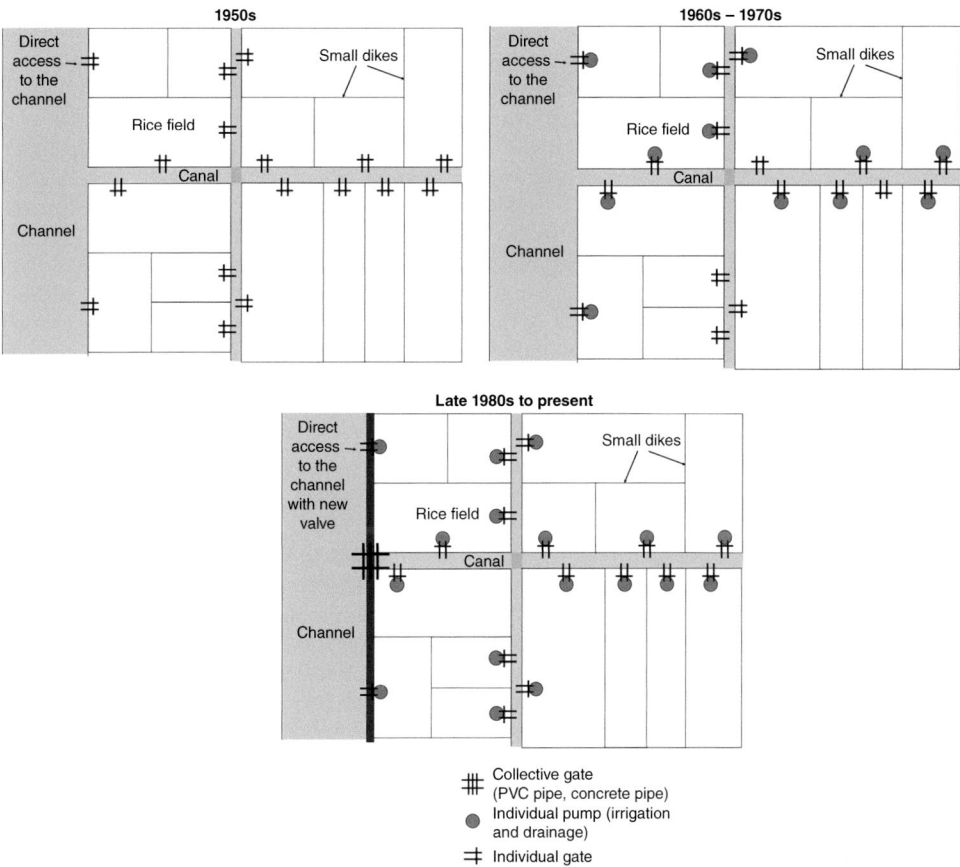

Fig. 4.21. Evolution of water management in the irrigated schemes of northern Thiện Trí commune (Champeaux, 2016, pp. 34, 42, and 48).

household) from absentee landlords living in the towns (Cái Bè) or cities (Mỹ Tho and Vĩnh Long, even Saigon). A quarter of the farmers were landless and worked as laborers for small direct landowners (10% of the population, 0.5–2 ha/ household). Land rent was very high.[20] More than half the families did not meet their needs with rice alone; they supplemented their income by selling their labor force. Only 15% of the farms, the largest (2–3 ha), had buffalos to be used for tilling. The others contracted out this service or, as in the smaller families' case, did the work manually. Apart from flood alluviation, rice fields were not fertilized; animal manure was reserved for gardens (Champeaux, 2016).

With the conflicts of the First Indochina War (1945–1954), and then the Vietnam War (1954–1975), insecurity set in throughout the country, and the commune of Thiện Trí was not spared. Because of the risks in collecting rents, large urban landlords released their capital by selling their land to small local landlords or to well-off farmers. The government of South Vietnam, under insistent pressure from the USA who financed the process, initiated two agrarian reforms. The aim was to promote direct land tenure in order to garner mass support of farmers,[21] potentially tempted by the communist propaganda on land reform in North Vietnam, by removing the burden of land rent.

The first reform (1955–1956) was timid[22] and only concerned with the redistribution of land above a 100 ha ceiling, while theoretically capping rents. As properties larger than 100 ha no longer existed in Thiện Trí, this had no effect on the commune (Dufay *et al.*, 1995). The second reform, more radical, was launched in 1970. Entitled *Land for Cultivators*, it included a ban on indirect tenancy and a 3-ha ownership ceiling. This reform was politically motivated by the war intensifying in the Delta and the difficulty of controlling the population. The villagers were grouped together at night along the national road in "strategic hamlets," guarded by the South Vietnamese army, and had difficulty accessing their remote fields during the day. In the commune, agrarian reform had led to the disappearance of tenant farming; social differentiation was reduced, with 75% of the population farming rice areas of 0.2–3 ha (50% with less than one hectare) under direct tenancy. However, landless villagers were excluded from

redistribution and continued to live on daily laborer wages.

At the same time, the South Vietnamese government subsidizes the dissemination of new technical packages: the first non-photoperiodic rice varieties with a short cycle and high yield potential from the Green Revolution, subsidized inputs, and individual motor pumps. The motor pumps facilitated drainage (in flood season) and irrigation (in dry season), allowing farmers to at least partially disconnect water management in the irrigated areas from the tidal flow and the water level in the channels (Fig. 4.21).

With non-photoperiodic varieties, the conditions united for the emergence and gradual spread of a second crop cycle over a large area of the floodplain (Figs. 4.17 and 4.22). Farmers were also using new short-cycle rice varieties to avoid cultivating during the water rise,[23] when the rice fields, always poorly diked, were exposed to flooding. With two harvests, annual production rose to 5 tons/ha, then gradually to 13–15 tons as yields increased with the widespread use of imported and subsidized chemical fertilizers (Fig. 4.23). The financing of the technological leap was possible with public subsidies and the disappearance of land rent (Le Coq, 2001).

On the Mekong natural levee, the former forest disappeared. It was progressively cleared and replaced by orchards that extended gradually to the north. With the new security of land tenure and increased means, farmers invested in orchards drained with furrows (Fig. 4.22 and Photo 4.3).

After the *de facto* reunification of Vietnam in April 1975, a new agrarian reform was initiated in the Delta. It aimed to redistribute land at a rate of 0.1 ha of rice fields per laborer. Its effect was limited in the commune of Thiện Trí, as only a quarter of landless villagers received plots to settle. While the Vietnamese economy was bled dry after more than 30 years of conflict, the government's priority was food security. Communist ideology was set aside by the new regime: there was no collectivization in the Mekong Delta (Le Coq, 2001; Tuan, 2007). The State invested in infrastructure to secure and increase rice production: the opening of canals and diking began in 1978 (Figs. 4.24 and 4.25). The dikes were built in collective construction work along the channels to raise the levees. They included gates that decoupled the primary channels

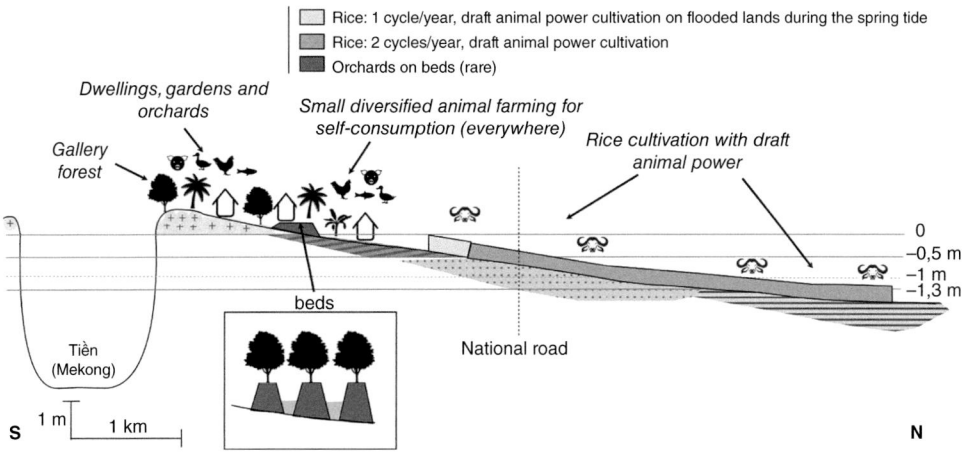

Fig. 4.22. South-north toposequence of Thiện Trí commune in the early 1970s (Champeaux, 2016, p. 30).

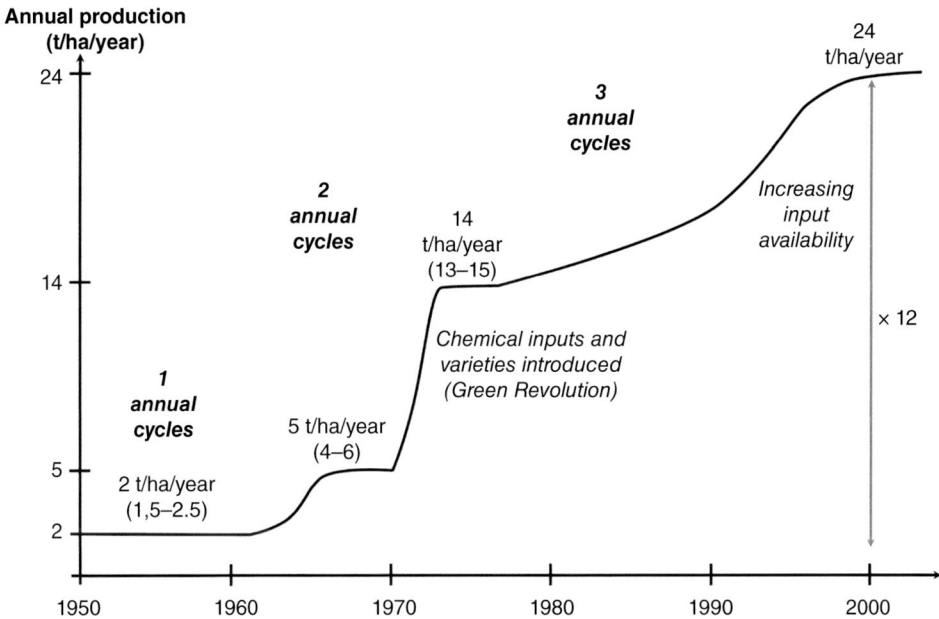

Fig. 4.23. Evolution of annual rice production in Thiện Trí commune (Champeaux, 2016, p. 66).

and canals from the secondary irrigation and drainage network (Figs. 4.22 and 4.24).

It was now possible to schedule a third rice cycle in the year, provided that the time spent on preparatory work for each cycle was reduced to a minimum. Moto-mechanized tillage developed with the introduction of multi-purpose power tillers (transport and plowing) and the first tractors, planned and subsidized by the State. The gradually generalized use of power tillers during the 1980s and 1990s was accompanied by the introduction of three annual rice-growing cycles. Due to the lack of grazing space (Fig. 4.25), buffalo farming declined until disappearing in the 2000s.

In the highest rice fields, it was only possible to grow one cycle of rice per year because of the lack of pump irrigation in the dry season when water is too far away (Figs. 4.22 and 4.25).

Fig. 4.24. Dikes in Thiện Trí commune (Champeaux, 2016).

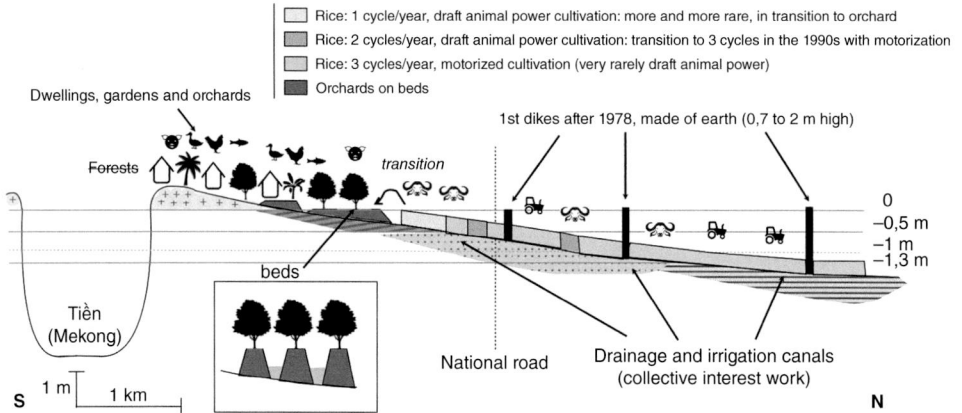

Fig. 4.25. South–north toposequence of Thiện Trí commune in the 1980s–1990s (Champeaux, 2016, p. 54).

Farmers were gradually converting these rice fields into orchards, providing increased household income and more continuous cash flow throughout the year.

Although it had theoretically been possible to obtain nearly 24 tons of rice per year[24] since the 1980s, the difficulties of supplying inputs via cooperatives limited results far from their potential. It was not until the gradual reorganization of private supply chains after the promotion of the market economy[25] that inputs, seeds, spare parts, services, loans, and traders were accessible to farmers in a timely manner. In 40 years, yearly rice yield had increased 12-fold (Fig. 4.23), while at the same time being gradually freed from the vagaries of flooding by increasingly State-led and State-financed diking. In the early 2000s, dikes were raised again and concreted, and new ones were built, including in the south to protect upland areas (Fig. 4.26).

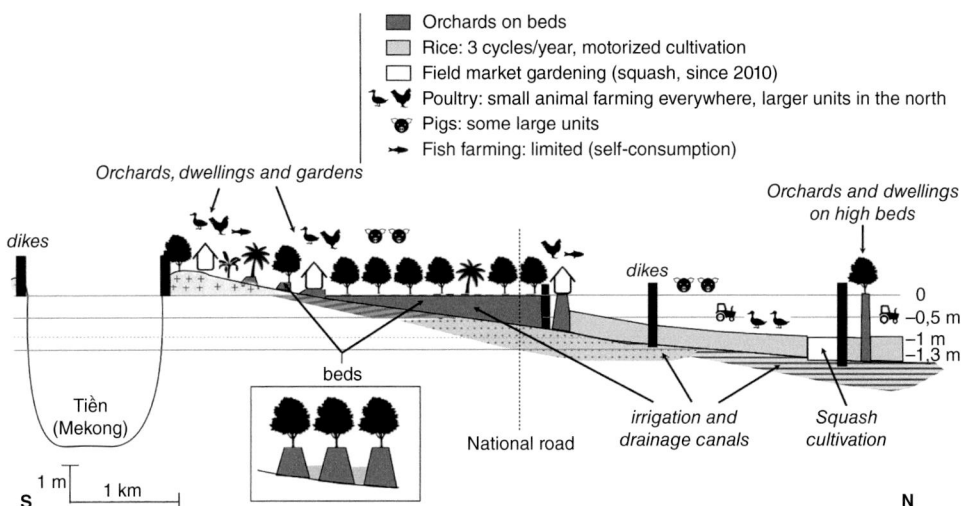

Fig. 4.26. South-north toposequence of the present-day commune of Thiện Trí (Champeaux, 2016, p. 64).

In the North, the widespread use of mo-to-mechanized tillage and harvesting generally involved using private service providers with tractors and combine harvesters (Photo 4.4). This allowed farmers to minimize periods of zero photosynthetic production in the fields, in favor of rice growth in three annual cycles.

The economic development of the mega-city of Ho Chi Minh City, and more generally Vietnam's urban areas, increased the demand for fresh produce. As the value-added per unit area of orchards is higher than that of rice fields, farmers had an interest in converting their farms, most of which were less than half a hectare. Diking in the southern half of the commune allowed farmers to drain the intermediate land and convert it into orchards (Figs. 4.26 and 4.27).

In this "socialist market economy," the State is very present with a constant priority of food security and sovereignty in rice. Direct subsidies are very limited (about 20 US $/ha), but the State intervenes indirectly through:

- financing and coordinating construction programs for irrigation, diking, and transport infrastructures (roads)
- subsidizing companies authorized to buy paddy rice from farmers, ensuring a remunerative and stable farmgate/field price for farmers, disconnected from world market variations (Chen and Saghaian, 2016)

- controlling exports via State-owned enterprises (50% of exports), volume quotas, and minimum export prices imposed on private enterprises grouped in the *Vietnam Food Association* (VFA)
- the near absence of import taxes on inputs and the liberalization of their marketing since the 1990s. The widespread use of fertilizers has been facilitated by the development of national production of nitrogen fertilizers from petroleum extracted from the China Sea since the early 2000s (50% of national demand)
- facilitating access to farming credits and equipment loans, at subsidized rates, granted by public banks
- the near absence of taxes for farm households

4.3.3 A thriving agricultural economy

In 50 years, agriculture in the Mekong Delta has been profoundly transformed. In a territory ravaged by 30 years of war, the local smallholders have made the region a major economic player in the national economy (68% of Vietnam's fruit production, General Statistics Office, 2016) and the global rice market (15% of world trade, Fig. 4.16). During this half-century metamorphosis, the State has accompanied the transformation of small-scale tenants, strangled by land rents

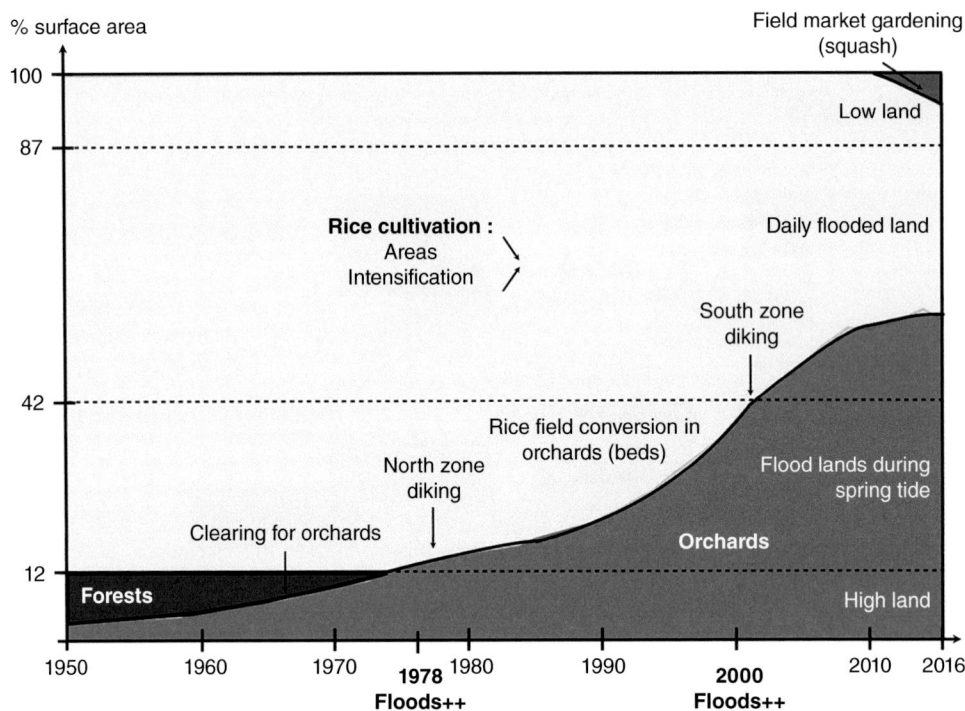

Fig. 4.27. Evolution of land use in Thiện Trí commune from 1950–2016 (Champeaux, 2016, p. 62).

and struggling to meet their basic needs, into successful commercial family agriculture.

The net value-added created by a laborer is significant, ranging from 1000–18,000 US$/year depending on the production system in Thiên Trí commune, compared to the income per capita in Vietnam that reaches 1500 US$ on average in 2015 (Champeaux, 2016). For most households, the income obtained is at least comparable to employment opportunities accessible in Ho Chi Minh City and Delta cities, thus limiting urban attraction and associated migration. However, the farms' small size[26] means that they must not be divided up during intergenerational succession if they are to remain viable. In each family, only the son who takes over the farm and his wife remain on the farm with his parents. The other brothers emigrate to cities or to mountainous agricultural regions with lower incomes (Déry, 2004; De Koninck and Rousseau, 2013). The sisters join their husband's economic units. Population growth is very low in rural areas of the Delta: 0.1% between 2010 and 2015,

compared to 8% for urban areas (General Statistics Office, 2016).

These substantial economic results are obtained at the cost of immense work on the family farm. The combination of many different tasks—planning and maintaining fields, maintaining dikes, the management sequence over three annual rice cycles, managing pig and poultry farms (parked), market gardening, marketing products, and supplying goods—overwhelms the annual labor calendar of all the farm's laborers and even involves recourse to the external workforce. These off-farm employees are left out of Delta development dynamics.

As descendants of mid-20th-century landless villagers, they have been ignored by the State, whether that of South Vietnam or the communist one, during successive agrarian reforms. From generation to generation, they lived on daily wage labor (300–900 US$/year/laborer; Champeaux, 2016). Many landless villagers have migrated to urban areas and this movement will continue in the future; they now account for less than 5% of households in the Delta.

4.3.4 A future exposed to global changes

Encouraged by the State for more than 50 years in the Mekong Delta, the Green Revolution has made possible the emergence of a prosperous family agriculture contributing to the country's sustained economic development. However, the policy options chosen expose farmers to risks caused by climate change and, more generally, global changes.

4.3.4.1 Climate change and marine transgression

The most publicized risk (Perrin *et al.*, 2016), as it is the most prominent, is the rise in the China Sea's average level due to its thermal expansion (MoNRE, 2009; Van Khiem, 2014). The Mekong Delta is the most threatened region in Southeast Asia (Nhat, 2015).

In 2009, researchers at the Vietnamese Ministry of Natural Resources and Environment (MoNRE) projected that the rise could reach 75 cm (medium scenario) in 2070 compared to the mean level in 2005, and as much as 100 cm (high scenario). This would translate, all else being equal, into transgressions of over 20–40% of the Delta land (MoNRE, 2009), which is no higher than 2.5 m. In 2019, a study reduced the Delta's predicted elevation to a mean of 0.8 m above sea level, which would submerge 75% of the Delta under a 1 m rise (Minderhoud *et al.*, 2019). In the study area province (Tiền Giang), this involved nearly one-third of the land area, including the lowest land north of Thiện Trí commune (Figs. 4.19 and 4.26). On an annual basis, climate change is modeling results in an increase of between 0.3 and 0.6 cm/year.

4.3.4.2 Climate change and flood risk

In the region, climate change is already reflected in a measurable rise in rainfall. The trend is set to increase in the rainy season in the Delta (MoNRE, 2009), as well as on the scale of the Mekong River watershed (Fig. 4.1) with higher interannual variability (Eastham *et al.*, 2008; Lacombe *et al.*, 2010; Icem, 2013a). This implies larger and more random floods (IPCC, 2012; Teng *et al.*, 2016), a phenomenon already notable in the 20th century (Delgado *et al.*, 2009). Exceptional floods, exceeding existing dikes' capacities, have marked farmers' memory with the damage's magnitude (1978 and 2000 in particular). If these exceptional hazards become increasingly frequent, deterioration risks hinder the development of the Delta's lowest areas, such as the north of the Thiện Trí commune, which could become uncultivated again as in the 1950s (Fig. 4.20).

4.3.4.3 Accelerated subsidence of the Mekong Delta

Due to alluvium compacting under its own weight, its porosity tends to decrease, reducing its thickness. Thus, the Delta "sinks" in relation to the sea level. In the Mekong Delta, this phenomenon tends to be reduced with the progressive diking of the land, which interrupts alluviation, the driving force behind subsidence.[27]

However, subsidence is also caused by freshwater withdrawals from groundwater tables, which are being drawn down by increased pumping. More than one million wells have been drilled in the Delta since the 1980s to cover domestic and industrial needs, with 430,000 m^3 of water withdrawn daily (Schmidt, 2015). With aquifer drawdown of up to 70 cm/year,[28] induced subsidence at the surface ranges from 1.5–5 cm/year. This effect of global change (Takagi *et al.*, 2016)[29] is in addition to marine transgression stemming from the China Sea's thermal expansion (climate change) at a rate seven times higher.

The relative rise in mean sea level is the absolute sum of thermal rise and subsidence: 75% of the Delta may be submerged by mid-century. Diking is the current main measure to counteract marine transgression. However, the phenomenon is continuous and tends to increase in speed and intensity. So, what will be the technological and budgetary limits of the "arms race" that is the recurrent raising of dikes?

4.3.4.4 Incidental salinization of water and soil

The eastern and southern parts of the Mekong Delta are bathed by saline surface waters due to tide penetration. The combination of subsidence and marine transgression also implies increased penetration of saline waters into the Delta, with an extension of saline areas and a westward upwelling of brackish areas (Smajgl *et al.*, 2015).

The salty and brackish areas' salinity levels are deleterious for most agricultural production, but favorable for aquaculture, especially shrimp farming. Although rice is a halotolerant plant, production is greatly reduced with increasing salinity. From an average of five tons of paddy rice per hectare per cycle for the Mekong Delta, the yield drops to three tons for tolerant varieties and less than one ton for sensitive varieties (Smajgl *et al.*, 2015).

Still, in the combination of the effects of global change beyond climate change alone, the multiplication of hydroelectric dams on the Mekong's tributaries and major course over the last 10 years tends to transfer part of the river's flow from the rainy season to the dry season. The reason for this is storage in reservoirs to produce energy continuously throughout the year. Partial suppression of the flood could limit its "flushing" effect in the Delta, which promotes the rise of dry-season marine inflows (Molle *et al.*, 2009; Matthews and Geheb, 2014). Hydropower development in the Mekong watershed would thus contribute to accentuating the effects of saltwater inflows into the Delta but would limit, to an extent yet to be determined, exceptional floods and their deleterious effects.

4.3.4.5   Other factors and effects of global change

In addition to the major causes of change discussed above, other factors are likely to interact, in probably smaller (or less region-specific) proportions:

- Rising temperatures constituent of climate change could reduce crop yields, particularly in rice (Teng *et al.*, 2016). While quantification of the phenomenon remains fragile and debated (IPCC, 2014a; Chapter 7, this volume), it is clear that prioritizing the selection of short-cycle rice varieties is problematic; in particular, when the temperature increase results in an accelerated vegetative cycle, reducing the plant's growth period and thus the yield's constitution.
- Erosion of rice agrobiodiversity in the Delta with the widespread adoption of a handful of Green Revolution varieties. For example, only one variety (IR504), selected for its high yield potential, is now grown on the thousand rice farms in the Thiện Trí commune. Vietnamese agronomic research has begun to work on obtaining varieties with more diversified characteristics (resistance to submergence, drought resistance, etc.), but there is still a long way to go before these can be offered to farmers.

- The development of hydroelectricity, already introduced, will result in a reduced sediment load downstream in the river, reducing alluvial deposits in the Delta. This will have an effect on subsidence and on the renewal of crop fertility. If the massive use of fertilizers covers the crops' needs, the soil's organic matter content is reduced and water pollution problems are major[30] (Sebesvari *et al.*, 2012).
- The evolution of the domestic market for fresh fruit and pork and world markets for rice, shrimp, and so on.

For all these factors, it is often difficult to isolate those specifically related to climate change because of other global change factors' synergistic or antagonistic effects, whether societal, technical, infrastructural, economic, and so on.

4.3.4.6   Prevention and adaptation policies

With the country's long maritime coastline and exposure to marine transgression in the Mekong Delta and, to a lesser extent, the Red River Delta, the Vietnamese government has been involved in defining policies to prevent the negative effects of climate change and to adapt the national economy since the early 2000s.

The measures were first institutional with the creation of an inter-ministerial government committee[31] responsible for policy design and coordination. The first measure was the publication of a decree by the Prime Minister[32] in 2008 concerning:

- the development of studies and research to identify and characterize the foreseeable effects of climate change;
- the institutional development necessary to take these effects into account; and
- raising awareness of climate change among institutions at all levels and among the population.

Within this institutional framework, research centers and universities (especially those in the deltas) have undertaken a multitude of research projects on climate change. Reports and publications are now abundant.

In a second step, the government prepared the "National Strategy on Climate Change" in 2011. In 2012, this strategy was translated into the National Action Plan for Responding to Climate Change 2012–2020, as well as a series of sector-specific texts including targets for reducing greenhouse gas emissions, protecting the environment (waste recycling and forest protection), and fighting poverty, as well as converting to a low-carbon, green economy.

This political voluntarism and profusion of legislative and executive texts have attracted many bilateral and multi-lateral official development assistance agencies[33] (Nhat, 2015). Unfortunately, concrete measures directly applicable to the benefit of Mekong Delta farmers have yet to be invented (Van Khiem, 2014).

The proposed project is to isolate the Delta from the sea with a series of giant dikes and gates on the river arms, similar to Dutch polders. This colossal and radical project would require 5–8 bn US$ of investment and more than 500 million US$ of annual maintenance (Smajgl et al., 2015). In addition to these amounts, more funding would be required in case the infrastructure is damaged or is insufficient to cope with the numerous typhoons affecting the east coast of Vietnam. The intensity of these typhoons may increase in the future.

The high degree of specialization of the various farms accentuates their sensitivity to hazards. As they only have access to a single ecosystem, farmers develop it to the best of their abilities. Nevertheless, they remain vulnerable if environmental or market conditions fall outside the safety zone built up over the course of history. For example, future exceptional floods, which are more frequent, could exceed the dikes' height and expose farmers in the lowest agricultural production systems if the water table continues not to be drained.

The submersion of a significant portion of the Delta will bring down rice production, replaced by aquaculture. The process has already begun in the coastal mangrove, where shrimp farming is quickly expanding (Duy et al., 2022). Highly profitable, the activity is, though, risky and requires some capital, leading the poorest farmers to rent their flooded land to urban investors. Intensive shrimp farming in mangroves has turned out in Thailand not to be sustainable, leaving a sterile soil over-polluted with antibiotics; the risk is high that shrimp farming would be only a transient trend (a pioneer front in the mangrove) for the Mekong Delta (Quan et al., 2022). Moreover, even if shrimp farming is a locally profitable alternative in submerged paddy fields, it cannot solve the staple food crisis that will occur in the main cities of the global south, when the cheapest rice in the world will be lacking.

4.3.4.7 Conclusion: impact of global changes on agriculture in Thiện Trí

In a region exposed to annual Mekong flooding, farmers and the State have learned to prevent the risk of flooding while making the best use of water and drained alluvium. The State has enabled hazard control and risk reduction through regular and increasing investment in diking and irrigation infrastructure. Successive agrarian reforms have allowed the emergence of efficient family farming. By maximizing value-added after investing in equipment (moto-mechanization) and plantations, farmers have made the Delta a prosperous region, a "rice granary" (and fruit) for the country. However, this success story is increasingly exposed to global change's new risks: marine transgression, accelerated subsidence, and increased risk of flooding.

4.4 Conclusion

With the political will in Cambodia and, even more so, in Vietnam, studies and research have multiplied to identify and characterize the effects of climate change in the Lower Mekong. While physical phenomena are extensively analyzed and modeled, the characterization of climate change's effects on farmers systematically suffers from the following shortcomings:

- Farmers are considered a homogeneous category of economic agents. Socioeconomic differentiation is not taken into account. Also ignored is unequal access to different ecosystems, on which the resilience, vulnerability, and adaptive capacities of different agricultural production systems nevertheless depend when facing the effects of global change.

- The projected effects of climate change are compared to a reference situation, which is the current baseline, without taking into account that the counterfactual (what would happen without climate change?) is also a dynamic, changing scenario.
- The effects of climate change are modeled "all things being equal" without considering interactions with other global change dynamics, or farmers' reactivity in the face of adversity.

The two regions studied are only 300 km apart in the Lower Mekong Valley. However, the agrarian systems differ radically in terms of agroecosystems, agricultural practices, social relations, policies, and historical dynamics of socioeconomic differentiation. This gives the different production systems varying levels of resilience and vulnerability, as well as adaptive capacities. Cambodian farmers seek to diversify their cropping systems by accessing different ecological stories to limit the effects of the uncontrolled hazard of flooding. As for those in the Mekong Delta, they have taken advantage of risk reduction through State-funded diking to specialize and increase family labor productivity. Adaptation to climate change, and more generally to global changes, does not rely on a single solution, but on targeted and carefully selected actions based on a detailed knowledge of local dynamics.

Notes

[1] In the Mekong Delta, the flow is 15,000 m^3/s on average, 150,000 m^3/s in October.

[2] 20,000 species of plants, 1200 species of birds, 430 species of mammals, 800 species of reptiles and amphibians, and 1700 species of fish have been identified in the Greater Mekong Region; 50% of the vascular plant species and 42% of the vertebrate species are endemic (source: IUCN).

[3] Intergovernmental Panel on Climate Change (IPCC).

[4] The choice of this site is much owed to Jean-Christophe Diépart, a researcher on land issues in Cambodia, and to André Pouillès-Duplaix, then director of the AFD's Cambodia-Laos agency. We would like to express our sincere thanks to them.

[5] Long-stemmed varieties of rice, whose tillers extend from the taproot as the water level slowly rises (the adjective "floating" is a misnomer but an accepted one). If the flood is rapid, the plants are drowned and lost.

[6] Stung Sen is a tributary of the Tonle Sap that flows through the town of Kampong Thom (Fig. 4.6).

[7] These are the only varieties from agronomic research (Irri) used in the region.

[8] For a detailed presentation of these different cropping systems, please refer to the dissertation by Lécuyer and Wiel (2014, pp. 71–100).

[9] 1.7–3.4 million, according to Heuveline (2001).

[10] Mobilized *manu militari* as a labor force just after the takeover for irrigation infrastructure projects, the villagers were unable to plant nurseries and transplant rice for the 1975 rainy season, resulting in rural famine in the months that followed.

[11] Literally, Krom Samaki means group of mutual aid, solidarity, and friendship.

[12] While the 1991 land titling reform has only been marginally implemented in the study area, land tenure is *de facto* secure and borders on private ownership (Diépart and Sem, 2014).

[13] This may contribute to their vulnerability to climate risk when workers have to postpone urgent activities in their fields for salaried tasks.

[14] These spaces are acquired in a coercive manner from the farmers who previously exploited these lands.

[15] Investors are also attracted by the ease of the "everything but Arms" initiative, which has allowed Cambodian exporters to access the European Union market duty-free since 2009 (Diépart and Sem, 2014).

[16] J.-C. Diépart, personal communication on research conducted in 2009 in Sravov, about ten kilometers east of the study area.

[17] Ho Chi Minh City had a population of 8.5 million in 2016.

[18] The choice of this site is much owed to Professors Nguyễn Ngoc Thùy, Director of International Relations, and Nguyen Van Ngai, Dean of the Faculty of Economics at Nông Lâm University in Ho Chi Minh City. We would like to express our sincere thanks to them.

[19] During high spring tides, the water can rise up to 1.6 m on average above the reference sea level.

[20] Land rent increases with surface area, from 650 kg of paddy/ha (25% of the average harvest) for a rented area of less than one hectare, to 1.5 tons/ha (65% of the average harvest) for an area of three hectares. Landlords leave villagers only the bare minimum to reproduce their family labor force. Rent increases with debt, especially in the years following poor harvests. It can lead to extreme dependence on owners (Champeaux, 2016).

[21] The political approach is comparable to the land reforms in Taiwan (1949–1951) and South Korea (1953–1955), which were also initiated and financed by the USA.

[22] The Gini index of land distribution in the Mekong Delta slightly decreased from 0.84 in 1955 to 0.80 in 1966 (Tuan, 2007).

[23] These short-strawed varieties do not tolerate flooding very well.

[24] Across the Mekong Delta, three-cycle rice cultivation remains limited. Double cropping dominates (Nguyen et al., 2015).

[25] The policy of 'socialist market economy (Đổi Mới)', initiated by the only party and the State in 1986, results in liberalization of the economy during the 1990s.

[26] Small family farms average 1.6 ha in size (Champeaux, 2016).

[27] Subsidence does not stop immediately with the cessation of alluviation. It continues for a few years, while tending to cancel itself out due to the inertia of compaction and the progressive mineralization of organic matter.

[28] Average aquifer drawdown was measured over the period 1995–2010 in 106 wells (Erban et al., 2014).

[29] The increase in groundwater withdrawals is due to a combination of industrial development in the region, urban population growth, and improvements in the populations' standard of living.

[30] Pollution is exacerbated by the equally massive use of pesticides. Surface water pollution leads to the domestic use of deep water, thus contributing to subsidence's acceleration.

[31] The secretariat of the interdepartmental government committee is housed in the MoNRE.

[32] Decision on approval of the National Target Program to respond to climate change. n°158/2008/QD-TTg, The Prime Minister, Hanoi December 12, 2008.

[33] Among these agencies, we note the presence of the World Bank, Asian Development Bank, UNDP, JICA, USAID, the French Development Agency, or the Dutch Cooperation.

References

Arias, M.E., Cochrane, T.A., Piman, T., Kummu, M., Caruso, B.S. and Killeen, T.J. (2012) Quantifying changes in flooding and habitats in the Tonle Sap Lake (Cambodia) caused by water infrastructure development and climate change in the Mekong Basin. *Journal of Environmental Management* 112, 53–66.

Champeaux, E. (2016) Research on dynamics of agrarian systems in the Mekong Delta: Thiện Trí, Tiền Giang, Vietnam. MSc thesis, UFR Agriculture comparée et Développement agricole. AgroParisTech, AFD, NLU, Paris, France, 120 pp.

Chen, B. and Saghaian, S. (2016) Market integration and price transmission in the world rice export markets. *Journal of Agricultural and Resource Economics* 41(3), 444–457.

De Koninck, R. and Rousseau, J.-F. (2013) Pourquoi et jusqu'où la fuite en avant des agricultures sud-est asiatiques? *L'Espace géographique* 42(2), 143–164.

Delgado, J.M., Apel, H. and Merz, B. (2009) Flood trends and variability in the Mekong River. *Hydrology and Earth System Sciences* 6(3), 6691–6719.

Déry, S. (2004) *La colonisation agricole au Viêt Nam*. Presses de l'Université du Québec, Québec, Canada, 310 pp.

Diépart, J.-C. and Sem, T. (2014) *Cambodian Peasants and the Formalization of Land Rights: Historical Overview and Current Issues*. AgroParisTech/CTF, Paris, France, 94 pp.

Doch, S., Diépart, J.-C. and Chinda, H. (2015) A multi-scale flood vulnerability assessment of agricultural production in the context of environmental change: The case of the Sangkae River watershed, Battambang province. In: Diépart, J.-C. (ed.) *Learning for Resilience*. The Learning Institute, Phnom Penh, Cambodia, pp. 19–50.

Dufay, A., Linh, N.D. and Veloso, A. (1995) Étude d'une région agricole du delta du Mékong. Mémoire de fin d'étude, UFR Agriculture Comparée et Développement Agricole. INAPG, Paris, France.

Dugan, P.J., Barlow, C., Agostinho, A.A., Baran, E., Cada, G.F. *et al.* (2010) Fish migration, dams, and loss of ecosystem services in the Mekong Basin. *Ambio* 39(4), 344–348.

Duy, D.T, Nga, N.H., Berg, H. and Da, C.T. (2022). Assessment of technical, economic, and allocative efficiencies of shrimp farming in the Mekong Delta, Vietnam. *Journal of the World Aquaculture Society* 54(4), 915–930. doi:10.1111/jwas.12915

Eastham, J., Mpelasoka, F., Mainuddin, M., Ticehurst, C., Dyce, P. *et al.* (2008) *Mekong River Basin Water Resources Assessment: Impacts of Climate Change.* CSIRO, Canberra, Australia, 153 pp.

Erban, L.E., Gorelick, S.M. and Zebker, H.A. (2014) Groundwater extraction, land subsidence, and sealevel rise in the Mekong Delta, Vietnam. *Environmental Research Letters* 9(8), 084010.

General Statistics Office (2016) *Statistical Yearbook of Vietnam 2015.* Statistical Publishing House, Hanoi, Vietnam.

Ham, K. and Someth, P. (2015) Impacts of social-ecological change on farming practices in the stung hrey-bak watershed, Kampong Chang, Cambodia. In: Diépart, J.-C. (ed.) *Learning for Resilience.* The Learning Institute, Phnom Penh, Cambodia, pp. 51–76.

Heuveline, P. (2001) The demographic analysis of mortality crises: the case of Cambodia, 1970–1979. In: Reed, H. and Keely, C. (eds) *Forced Migration and Mortality.* National Academy Press, Washington, DC, pp. 102–129.

Hoanh, C.T., Jirayoot, K., Lacombe, G. and Srinetr, V. (2010a) Comparison of climate change impacts and development effects on future Mekong flow regime. In: *Proceedings of the Conference Modelling for Environment's Sake, International Congress on Environmental Modelling and Software, Fifth Biennial Meeting.* Ottawa, Ontario, Canada, 9 pp.

Hoanh, C.T., Jirayoot, K., Lacombe, G. and Srinetr, V. (2010b) *Impacts of Climate Change and Development on Mekong Flow Regime, First Assessment.* MRC Technical Paper 29, Mekong River Commission, Vientiane, Laos, 104 pp.

Icem (2013a) *Mekong ARCC Climate Change Impact and Adaptation Study for the Lower Mekong Basin: Main Report.* USAID, Bangkok, Thailand, 294 pp.

Icem (2013b) *Mekong ARCC Mekong Adaptation and Resilience to Climate Change: Synthesis Report.* USAID, Bangkok, Thailand, 233 pp.

Icem (2014a) *Mekong ARCC Climate Change Impact and Adaptation on Agriculture.* USAID, Bangkok, Thailand, 228 pp.

Icem (2014b) *Mekong ARCC Climate Change Impact and Adaptation on Fisheries.* USAID, Bangkok, Thailand, 159 pp.

Icem (2014c) *Mekong ARCC Climate Change Impact and Adaptation on Livestock.* USAID, Bangkok, Thailand, 160 pp.

Icem (2014d) *Mekong ARCC Climate Change in the Lower Mekong Basin: An Analysis of Economic Values at Risk.* USAID, Bangkok, Thailand, 38 pp.

IPCC (2012) *Managing the Risks of Extreme Events and Disasters to Advance Climate Change Adaptation. A Special Report of Working Groups I and II of the Intergovernmental Panel on Climate Change.* Field, C.B., Barros, V., Stocker, T.F., Qin, D., Dokken, D.J. *et al.* (eds) Cambridge University Press, Cambridge, UK, 582 pp.

IPCC (2014a) *Climate Change 2014: Impacts, Adaptation, and Vulnerability. Part A: Global and Sectoral Aspects. Contribution of Working Group II to the Fifth Assessment Report of the Intergovernmental Panel on Climate Change.* Field, C.B., Barros, V.R., Dokken, D.J., Mach, K.J., Mastrandrea, M.D. *et al.* (eds) Cambridge University Press, Cambridge, UK, 1132 pp.

IPCC (2014b) *Climate Change 2014: Impacts, Adaptation, and Vulnerability. Part B: Regional Aspects. Contribution of Working Group II to the Fifth Assessment Report of the Intergovernmental Panel on Climate Change.* Barros, V.R., Field, C.B., Dokken, D.J., Mastrandrea, M.D., Mach, K.J. *et al.* (eds) Cambridge University Press, Cambridge, UK, 688 pp.

Johnston, R., Lacombe, G., Hoanh, C.T., Noble, A., Pavelic, P. *et al.* (2010) *Climate Change, Water and Agriculture in the Greater Mekong Subregion.* IWMI Research Report 136, International Water Management Institute, Colombo, Sri Lanka, 60 pp.

Kreft, S., Eckstein, D., Junghans, L., Kerestan, C. and Hagen, U. (2015) *Global Climate Risk Index 2015: Who Suffers Most From Extreme Weather Events? Weather-Related Loss Events in 2013 and 1994 to 2013.* Briefing paper. German Watch, Bonn, Germany, 32 pp.

Kummu, M., Tes, S., Yin, S., Adamson, P., Jozsa, J. *et al.* (2014) Water balance analysis for the Tonle Sap lake-floodplain system. *Hydrological Processes* 28, 1722–1733.

Lacombe, G., Hoanh, C.T. and Smakhtin, V. (2010) Multi-year variability or unidirectional trends? Mapping long-term precipitation and temperature changes in continental Southeast Asia using PRECIS regional climate model. *Climatic Change* 113(2), 285–299.

Lacombe, G., Smakhtin, V. and Hoanh, C.T. (2013) Wetting tendency in the central Mekong basin consistent with climate change-induced atmospheric disturbances already observed in East-Asia. *Theoretical and Applied Climatology* 111(1–2), 251–263.

Le Coq, J.-F. (2001) Libéralisation économique au Viêt-Nam, intensification rizicole et diversification: étude du cas de la région d'Ô Môn (delta du Mékong). Thèse de doctorat, Chaire d'Agriculture comparée et Développement agricole, INA-P G, Paris, France.

Lécuyer, C. and Wiel, F.-A. (2014) *Diagnostic Agraire, plaine rizicole de Kampong Thom, Cambodge: Adaptation des pratiques agricoles au changement climatique*. Mémoire de fin d'étude, UFR Agriculture comparée et Développement agricole. AgroParisTech, AFD, Paris, France, 179 pp.

Matsui, S., Keskinen, M., Sokhem, P. and Nakamura, M. (2006) *Tonle Sap: Experience and Lessons Learned Brief*. Global Environment Facility, Washington D.C., 14 pp. Available at: https://www.iwlearn.net/documents/33349 (accessed 9 October 2024).

Matthews, N. and Geheb, K. (2014) *Hydropower Development in the Mekong Region: Political, Socio-Economic and Environmental Perspectives*. Routledge, London, 218 pp.

Minderhoud, P.S.J., Coumou, L., Erkens, G., Middelkoop, H. and Stouthamer, E. (2019) Mekong delta much lower than previously assumed in sea-level rise impact assessments. *Nature Communications* 10(1): 3847.

Molle, F., Foran, T. and Kakonen, M. (eds) (2009) *Contested Waterscapes in the Mekong Region: Hydropower, Livelihoods and Governance*. Earthscan, London, 416 pp.

MoNRE (2009) *Climate Change, Sea Level Rise Scenarios for Vietnam*. Ministry of Natural Resources and Environment, Hanoi, Vietnam, 34 pp.

MRC (2005) *Overview of the Hydrology of the Mekong Basin*. Mekong River Commission, Vientiane, Laos, 73 pp.

Murphy, T.I., Irvine, K. and Sampson, M.K. (2013) The stress of climate change on water management in Cambodia with a focus on rice production. *Climate and Development* 5(1), 77–92.

Nguyen, D.B., Clauss, K., Cao, S., Naeimi, V., Kuenzer, C. and Wagner, W. (2015) Mapping rice seasonality in the Mekong delta with multi-year envisat ASAR WSM data. *Remote Sensing* 7(12), 15868–15893.

Nhat, L.M. (2015) *Climate Change Impacts and Adaptation Efforts in Vietnam*. Communication of the Ministry of Natural Resources and Environment. MoNRE, Hanoï, Vietnam, 25 pp.

Nuorteva, P., Keskinen M. and Varis, O. (2010) Water, livelihoods and climate change adaptation in the Tonle Sap lake area, Cambodia: learning from the past to understand the future. *Journal of Water and Climate Change* 1(1), 87–101.

Perrin, C., Aïchi, L. and Giraud, E. (2016) Climat: vers un dérèglement géopolitique? Rapport d'information n°14 (2015–2016) de la Commission des affaires étrangères, de la défense et des forces armées du Sénat. Available at: www.senat.fr/rap/r15-014/r15-0141.pdf (accessed 7 August 2024).

Pillot, D. (2007) *Jardins et rizières du Cambodge, les enjeux du développement agricole*. Karthala, Paris, France, 522 pp.

Quan, T.L., Tuan, V.A., Thuy, N.T.B., Huynh, L.D. and Duc, N.M. (2022) A closer look into shrimp yields and mangrove coverage ratio in integrated mangrove-shrimp farming systems in Ca Mau, Vietnam. *Aquaculture International* 30(2), 863–882.

Roth, C.H. and Grunbuhel, C.M. (2012) Developing multi-scale adaptation strategies: a case study for farming communities in Cambodia and Laos. *Asian Journal of Environment and Disaster Risk Management* 4(4), 441–462.

Schmidt, C. (2015). Alarm over a sinking delta. *Science* 348(6237), 845–846.

Sebesvari, Z., Le Huong, T.T., Pham, V.T., Arnold, U. and Renaud, F.G. (2012) Agriculture and water quality in the Vietnamese Mekong delta. In: Renaud, F. and Kuenzer, C. (eds) *The Mekong Delta System: Interdisciplinary Analyses of a River Delta*. Springer, London, pp. 331–361.

Smajgl, A., Toan, T.Q., Nhan, D.K., Ward, J., Trung, N.H. *et al.* (2015) Responding to rising sea levels in the Mekong delta. *Nature Climate Change* 5, 167–174.

Takagi, H., Thao, N.D. and Anh, L.T. (2016) Sea-level rise and land subsidence: impacts on flood projections for the Mekong Delta's largest city. *Sustainability* 8(9), 959–973.

Teng, P.S., Caballero-Anthony, M. and Anderias Lassa, J. (2016) *The Future of Rice Security under Climate Change*. Centre for Non-Traditional Security Studies (NTS), Singapore, 81 pp.

Tuan, D.T. (2007) Consecutive agrarian reforms and success in family farming. In: Merlet, M. (ed.) *Land Policies and Agrarian Reforms*. Tome 2/2. Agter, Paris, pp. 15–18.

Van Khiem, M. (2014) *Vietnam National Strategy on Climate Change*. Communication à la conférence 2nd Mekong Climate Change Forum. MRC, Siem Reap, Cambodia, 25 pp.

Plate 1

Photo 1.1. The use of draught power in the peanut basin, a key factor in adaptation to drought (Le Goff, 2016).

Photo 1.2. Ruminant enclosures, one of the adaptations facilitating fertility transfers in the peanut basin (Garambois, 2016).

Photo 1.3. Small irrigated perimeter fed by wells for off-season market gardening, crops in the peanut basin (Le Goff, 2016).

Plate 2

Photo 2.1. Chitemene field in Northern Zambia. Left: pyre formed in the center of the field for burning. Right: first-year finger millet, ready for harvesting (finger millet in the center, pruned halo, but not cultivated on the left) (Photos by C. Rémy on the left, H. Cochet on the right 2014).

Photo 2.2. Small irrigated perimeter from spring (center) and lowland fields cultivated with raised beds and mounds (right) in the southern highlands of Tanzania (Courtesy of Google Earth).

Photo 2.3. The importance of small-scale irrigation equipment: a treadle pump in a small, irrigated perimeter in the southern highlands of Tanzania (Photo: N. Verhoog, 2014).

Plate 3

Photo 2.4. Isupilo small-scale irrigation perimeter in the Kiponzelo region (southern Tanzania) (Photo: N. Verhoog, Google Earth background).

Photo 2.5. Water circulation and small equipment (pipe) in the Isupilo irrigation perimeter (southern Tanzania) (Photos: H. Cochet, 2014).

Plate 4

Photo 3.1. Corn/rice association cropping and hazard management in the Lower Rufiji flood plain, southern Tanzania. (Photo by H. Cochet 5/2014.)

Photo 3.2. Catch-up transplanting in the lower Rufiji Valley, southern Tanzania. (Photo by H. Cochet 5/2014.)

Photo 3.3. Complete flooding of sandbanks and catch-up transplanting in the lower Rufiji Valley, southern Tanzania. The lower part of the field, freshly transplanted, is in the foreground; the part that resisted the flood is in the background (behind the field box). (Photo by H. Cochet 5/2014.)

Plate 5

Photo 4.1. School children in a fishing village on Tonle Sap Lake, Cambodia (Ducourtieux, 2014).

Photo 4.2. Irrigated rice fields in the high plain on the banks of the Tonle Sap, Kampong Thom, Cambodia (Ducourtieux, 2014).

Photo 4.3. Combine harvester in the rice fields of Thiện Trí, Mekong Delta, Vietnam (Ducourtieux, 2016).

Plate 6

Photo 4.4. Manual leveling for orchard drainage (rice field conversion) in Thiện Trí, Mekong Delta, Vietnam (Champeaux, 2016).

Photo 5.1. Rice cultivation on large irrigated perimeters in the Senegal River delta. Senegal (here localized use of herbicides using a backpack sprayer) (Garambois, 2016).

Plate 7

Photo 5.2. Small plots belonging to family farmers in contact with the booming agribusiness in the lower delta (here large plots equipped with irrigation pivots) (Google Earth).

Photo 5.3. Installation of irrigated perimeters on the dune lands of the lower delta (here as part of the PDMAS) (Garambois, 2016).

Plate 8

Photo 6.1. Staircased planned hillside below Nyandira village, Uluguru Mountains, Tanzania. (Photo: H. Cochet 5/2016.)

Photo 6.2. Hosing beans from a primary channel in the Uluguru Mountains, Tanzania. The primary channel is visible at the top left of the picture, at the upper edge of the field. (Photo: H. Cochet 5/2016.)

Plate 9

Photo 6.3. Pigsty in the immediate vicinity of a house in the Uluguru Mountains, Tanzania. Below, the banana plantation benefits from manure runoff. (Photo: H. Cochet 5/2016.)

Photo 6.4. The technicist myth of the radical terrace: increased erosion and abandonment, as demonstrated by this "planned" hillside below Tchenzema parish (Uluguru Mountains, Tanzania). (Photo: H. Cochet 5/2016.)

Plate 10

Photo 6.5. A staircased planned hillside below a hamlet in Tchenzema, Uluguru Mountains, Tanzania. (Photo: H. Cochet.)

Photo 6.6. Ridge plowing on a moderate slope in the Uluguru Mountains (Tanzania): flat part at the bottom of the field and perpendicularly oriented ridges. (Photo: H. Cochet 5/2016.)

Plate 11

Photo 6.7. Step plowing on the steep slopes of the Uluguru Mountains (Tanzania): the biomass windrow swath, and in particular its unrooted part, is exposed at the base of the ridge. Taro and banana plants are maintained and contribute to reinforce the whole system. (Photo: H. Cochet 5/2016.)

Photo 6.8. Ridge plowing on steep slope in the Uluguru Mountains. The secondary irrigation canal, with a stone edge, is visible on the left side of the picture. (Photo: H. Cochet 5/2016.)

Plate 12

Photo 6.9. Vegetable terraces in the Uluguru Mountains (Tanzania): water from the secondary canal at the edge of the field is introduced to each terrace through a gutter at the foot of the bank. (Photo: H. Cochet 5/2016.)

Photo 6.10. Biomass burning on radical terraces, followed by hoe plowing (Uluguru Mountains, Tanzania). (Photo: H. Cochet.)

5 Drought, Hydraulic Planning and Development Models in the Sahel: The Case of the Senegal River Delta

Nadège Garambois[1]*, Samir El Ouaamari[1]*, Mathilde Fert[2], Léa Radzik[2], and Thibault Labetoulle[2]

[1]*Comparative Agriculture Training & Research Unit/UMR Prodig, AgroParisTech, Palaiseau, France;* [2]*Agroeconomist, AgroParisTech, Palaiseau, France*

5.1 Introduction

The Senegal River Delta region (Fig. 5.1) was already experiencing low rainfall totals before the 1970s–1980s drought episode that affected the whole Sahel. It was particularly prone to inter-annual variations in rainfall and river flooding, making rainfed agriculture hazardous (Lericollais, 1975) and farmers' and herders' ancient, systemic adaptation to random and occasionally extreme climatic and hydrographic conditions central. The series of extreme rainfall and river flooding deficits experienced by the Middle and Lower River Valley from the 1970s onward have, as in the peanut basin (see Chapter 1, this volume), profoundly weakened agriculture in this region.

However, the Senegal Valley now provides the bulk of national rice production (83% in 2012; MAER, 2014), a production unknown in the region before the 1960s and the result of an extensive hydraulic planning policy over the last 50 years. Largely supported by international donors, this policy was initiated during the colonial period but saw its first major developments, in particular the installation of dams up and downstream on the river, during the severe drought of the 1970s and 1980s. The delta region (Fig. 5.2), initially sparsely populated, is the area whose agriculture has been the most disrupted by this policy on developing irrigation. At the end of the 2000s, it alone accounted for half of the entire valley's irrigation surface area (Dahou, 2009) while still showing a strong potential for developing irrigation today, especially in the Lower Delta.

Initially conceived as a policy for adapting to climate change, such irrigation development is now increasingly seen as part of a food self-sufficiency strategy. However, despite the majority (60%) of public investments allocated to the agricultural sector being concentrated in hydro-agricultural planning, which is reinforced by growing private investments, rice imports remain colossal in Senegal, making up two-thirds of the rice consumed in 2014 (FAO) and 8.5% of the total value of imports (ANSD, 2008). This paper examines the role played by this planning

*Corresponding authors: nadege.garambois@agroparistech.fr; samir.elouaamari@agroparistech.fr

©2024 CAB International. *Agrarian Systems and Climate Change: Journeys of adaptation in the Global South* (eds H. Cochet, O. Ducourtieux, and N. Garambois)
DOI: 10.1079/9781800628137.0005

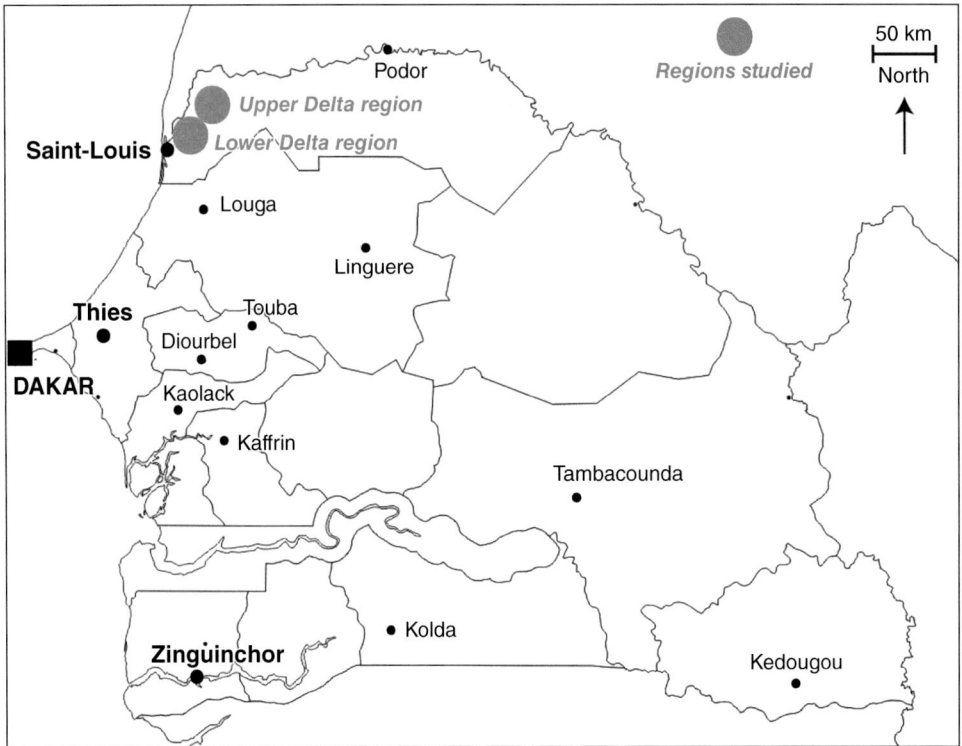

Fig. 5.1. Location of the Lower and Upper Delta regions of the Senegal River.

in the adaptation of agriculture in the Senegal River Delta to the profound and brutal climatic upheaval to which farmers have been subjected since the 1970s, and its economic, social, food, and environmental effects, as well as the development models this has brought about, up to the latest wave of irrigation development marked by the rise of commercial agriculture.

5.2 Agriculture in the Senegal River Delta and the Severe Drought of the 1970s and 1980s: Pushing Farmers' Adaptation Limits?

5.2.1 Agriculture in a Sahelian climate from the early 20th century, subject to a double hazard: rainfall and river flooding

Located at the northern end of Senegal, the Senegal River Delta region has recorded even lower rainfall than the northern part of the peanut basin since measurements were first taken (1892 for the Saint-Louis station) (see Chapter 1, this volume). In contrast to the peanut basin, this region has already been under a Sahelian climate (cumulative annual rainfall between 250 and 400 mm) since the first half of the 20th century (Fig. 5.3), with the 400 mm rainfall limit also being the supposed threshold for the existence of rainfed agriculture, and subject to strong interannual variability.

Before the hydraulic development of the latter half of the 20th century, which disrupted delta ecosystems, the region's landscape was structured around the Senegal River and its main distributaries (Djeuss, Lampsar) and organized between three main agroecological zones unequally subjected to their floods each year. The floods varied in magnitude but were differentiated by their topographic position: (i) in the low position, the "basins" were flooded for the longest amount of time; (ii) in the intermediate position, the Upper Delta levees and fluvial-deltaic

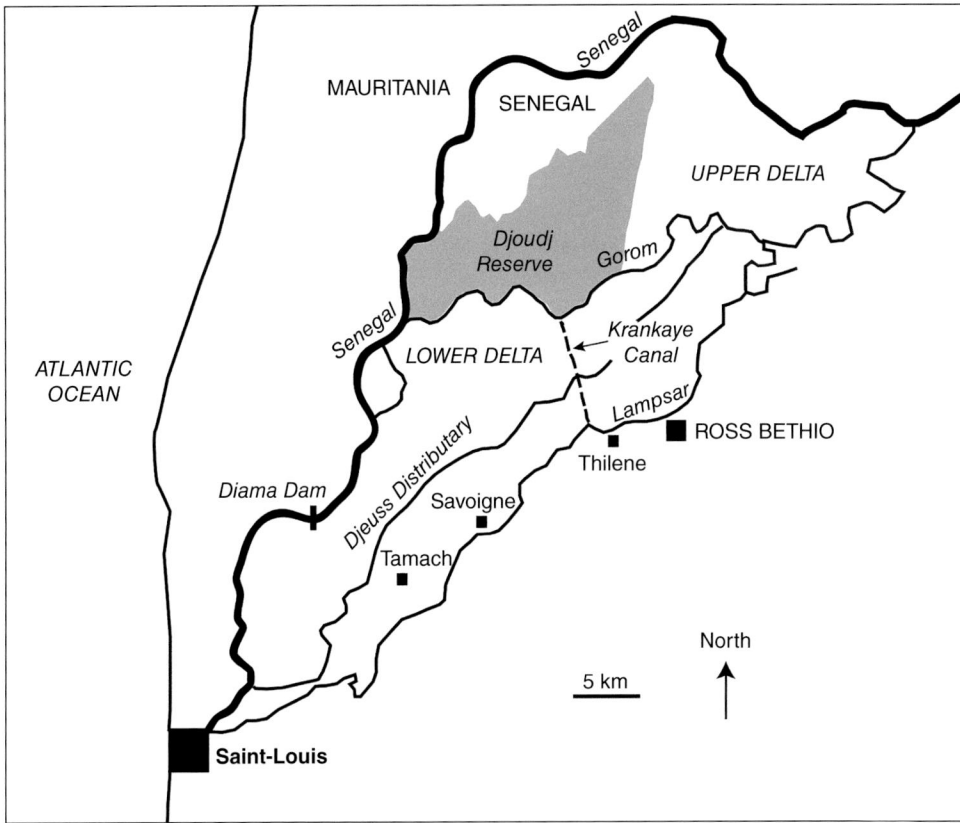

Fig. 5.2. Diagrammed organization of the hydrographic network and main hydraulic axes in the Senegal River Delta region. (From: Authors.)

deposits or the marine terraces and mudflats in the Lower Delta (which, over its geological history, is more subject to the influence of marine transgressions) were flooded for less time; and (iii) the fixed dunes were never flooded. The flood recession took place gradually from October to January, from the higher altitudinal positions to the lower ones. During the low-water period, saltwater flowed in the opposite direction, from the river estuary into the minor bed and on to Richard Toll, flowing back into the draining marigots and, in the dry season, stagnating in the lower parts of the basins. Like rainfall, the flood was and still is characterized by a strong hazard, both in intensity (water height) and in arrival date. At the time, the region's soils were marked by their halomorphic character, depending on the duration of saltwater intrusion, as well as the length and intensity of the rainy and dry seasons (Michel and Sall, 1984).

5.2.2 Sahelian agriculture has long been adapted to hazards

Faced with the vagaries of rainfall and the Sahelian climate, from the first half of the 20th century, Sahelian agriculture was only partially based on rainfed crops. Farmers practiced a typical rotation (millet/cowpea/watermelon/peanut/3–5 years of fallow land) on wooded parkland on the dunes in the rainy season. These rainfed crops, with uncertain harvests in low rainfall years, were complemented by flood recession crops, which played a key role in this agriculture. Cassava, corn, sweet potato, and market gardening crops (chili, African eggplant, tomato, etc.) were planted as the water level subsided, on areas that were covered by the river's flood during the rainy season, with the exception of the basins, where, in the dry season, saltwater flows back in and soils were too salt affected to be cultivated. The regular supply of alluvial deposits from the river's flooding and the floodplain areas' inundation allowed for cultivation every year. Cowpea and millet, sown in the rainy season (June), were harvested in September and October, respectively, and provided food for families until recession crops were harvested in February.

In the wet season, ruminant herds were driven to the dunes' wooded savannahs, behind the rainy season camps, where they consumed the grass layer present. They were then led to common grazing in rainfed crop fields as soon as the harvest was over and then to flooded areas during the rainy season (apart from the fraction reserved for flood recession crops), which were gradually cleared as the flood receded and where an herbaceous layer progressively developed. At the height of the dry season, leaves and fruit from trees (from the dunes' wooded parkland) supplemented the diet if grass was lacking on the floodplain. The night-time resting of animals on fields that were cultivated the following year ensured fertility transfer. The delta region was also a transhumance area for herders from the northern Ferlo or Mauritania who came to graze their herds. Fishing during the dry season, done every 2–3 days, took over from curdled milk to provide protein in families' diets at this time of year.

The semi-sedentary habitat followed alternating use of the different agroecological areas during the year, for both crops and cattle breeding. Villages were thus established on the dunes in the rainy season and on the edge of the recession areas in the dry season. At this time, land was not the limiting factor for this strictly manual agriculture. The area cultivated per agricultural worker was, however, limited by labor peaks, especially associated with weeding. Each worker could therefore cultivate about 0.5 ha of rainfed dunes and 0.2 ha of floodplain crops each year.

Social differentiation was above all based on herd size (which could lead farms to a certain degree of specialization) and villages' location in relation to the Senegal River and its branches (which determined both access to fish resources and proximity to the Mauritanian border for cross-border trade). With a herd of a few dozen head of cattle, families combined cropping with breeding and fishing. Families with large herds (more than 100 and up to 300 cows) specialized

in livestock. Those with only a few cows were settled at the foot of the dunes all year round, on the banks of the river or its branches. They combined rainfed crops, flood recession crops, and fishing in both the wet and dry seasons, entrusting their animals to dune village shepherds, who drove larger herds. Those who did not own their own pirogue had to pay part of their fishing earnings to the owner who lent them the boat and fell back on activities that required labor but not capital (charcoal production, mat weaving, etc.).

Without controlling natural phenomena (river flooding, rainfall), farmers were able to finely adapt their practices to this short rainy season as well as hydric and rainfall hazards. Following Lericollais (1975), our surveys show that this agriculture constituted a complex risk mitigation system based on a combination of productions and activities (cultivation, breeding, fishing, gathering, etc.), made possible by the complementary use of the different ecological areas throughout the year (rainfed and flood crops, progressive sowing of flood crops as the water recedes), the diversity of production (cereals, tubers, pulses, vegetables, livestock products, fishery products, fuel, etc.), and the food calendar.

Moving semi-sedentary camps between the rainy and dry seasons to be as close as possible to resources, the constitution of security food stocks (granaries, animal product storage techniques [meat that is dried, ground into powder, or cooked and preserved in fat]), bartering between producers with varying degrees of specialization, capital held in livestock reserves, and other activities deployed according to the capital available (charcoal, handicrafts, trade, etc.)—all these factors contributed to the robustness of this agrarian system against rainfall, water, and economic hazards, at a time when population density was low.

5.2.3 The limits of the Delta's ancient agrarian system in adapting to the drought of the 1970s and 80s

As in the peanut basin (see Chapter 1, this volume), the Senegal River Delta region has experienced a clear rainfall deterioration since the early 1970s, which was also observed throughout the valley and led to a massive, widespread reduction in the river's rise, contributing to a double disruption of agriculture in the region.

The description by Lericollais (1976) of the 1972 rainfall and water conditions in the Senegal Valley is edifying: annual rainfall dropped to 152 mm at the Saint-Louis station, the average flow of the Senegal River was reduced threefold, flooded areas were extremely small, and low water levels were reached very early, resulting in a strong influx of sea and saltwater in the lower Senegal River Valley. The harvest of flood recession crops did not reach 10% of the average production and rainfed crops were nil for millet and peanuts and very limited for cowpeas. The disruption of fish reproduction during the rainy season and the development and transport of fish larvae in floodwaters led to a halving of the fishing volume that year. The very low rainfall in the delta did not allow for the reconstitution of the dune rangelands' pasture swards, while, as in all of the northern Ferlo, the tree stratum was severely affected, when its fruits were essential resources: source of foreign exchange (Arabic gum), charcoal-making activities, fodder resource in the dry season, source of domestic wood, etc. (Poupon, 1976). While tree cutting, particularly for charcoal production, had already contributed to the gradual reduction of tree cover in the region, the prolonged drought greatly accentuated soil denudation, favoring wind morpho-dynamics and shaping the dunes (Roquet, 2008).

However, this was only the first year of a long series of deficits until the early 1990s (Fig. 5.3): the average annual rainfall between 1968 and 1992 was 223 mm, which is more than a third less than the 1946–1967 period, when average rainfall was already low compared to that recorded in previous decades. During this period, the region thus fell below the 250 mm limit that separates the Sahelian climate from the Sahelo-Saharan climate. At the same time, the intra-annual distribution of rainfall changed (Fig. 5.4): the useful rainfall of the four rainy season months (July to October) was reduced by an average of a quarter at the season's beginning and end (July and October) between the 1922–1945 and 1968–1992 periods, and by almost 50% at the height of the rainy season (August–September); the rare rainy episodes of the dry season were, on the contrary, more abundant during this long dry phase, which contributed to disrupting the metabolism of tree and shrub species through false starts in vegetation.

In the past, farmers were able to adapt to this hazard by sowing smaller areas with flood recession crops in years when the flood was lower (Lericollais, 1976) and occasionally compensated for lower harvests of rainfed crops in years with low rainfall with other productions or activities. However, these major, repeated rainfall and flood deficits caused significant problems to this agriculture. The crisis affected

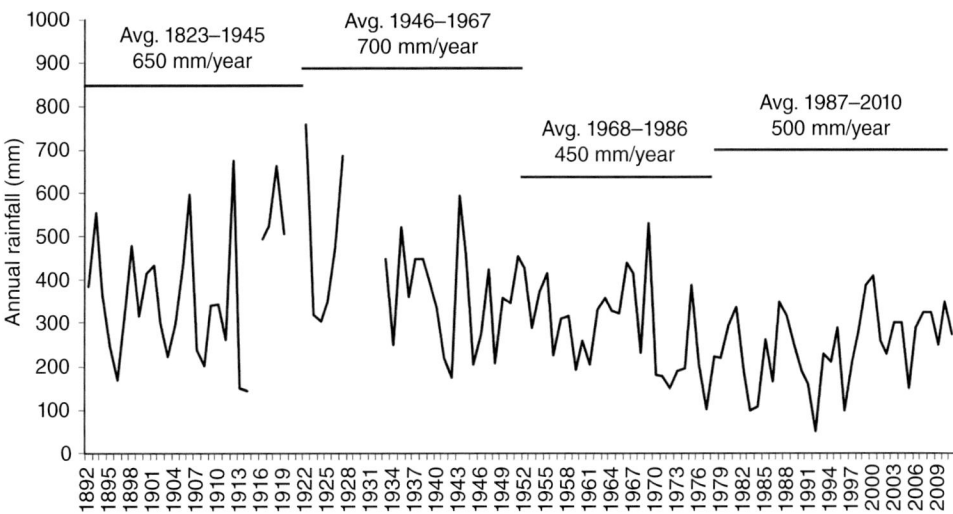

Fig. 5.3. Evolution of annual rainfall in Saint-Louis (1892–2011). (Courtesy of: Le Borgne (1988); Kamara (2013); graph: authors.)

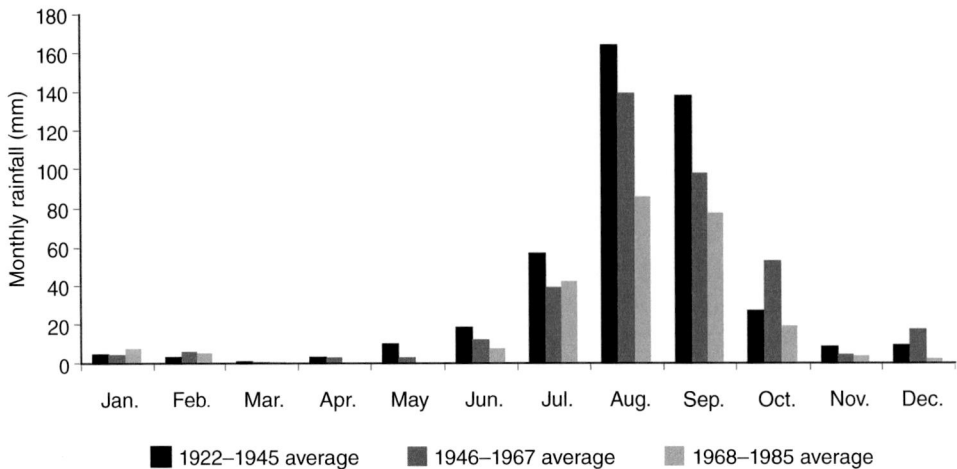

Fig. 5.4. Average monthly rainfall in Saint-Louis. (From: Le Borgne, 1988; graph: authors.)

all components of the agrarian system: very poor grass and shrub growth in the steppe (which eventually reached the tree cover itself and did not allow for sufficient feeding of the herds), low or no rainfed crop yields, strong reduction of the areas sown with flood recession crops and low yields, depletion of fish resources, etc. Families' survival during this period often depended on emergency food aid (Reboul, 1984). Livestock decapitalization was brutal: Lericollais (1976) indicates a three-quarter decrease in the number of animals in 1972 in certain sectors of the Valley, and the Livestock Services of the Dagana sector, which covers the Lower River Valley, registered a threefold reduction in the number of animals (deaths or distress sales at low prices), for which larger movements by transhumance were not enough to compensate.

5.3 Development Policy in the Delta until the 1990s: Damming the Climate Crisis?

5.3.1 Accelerated irrigation development with the climate crisis: security through total water control and extension of irrigated areas

In order to try to replace rainfed and flood recession crops with irrigated crops, the Senegalese government accelerated and reinforced the development of large rice-growing areas initiated in the delta in the 1950s and 1960s under the aegis of the *Société d'Aménagement et d'Exploitation des terres du Delta* (Senegal River Delta Land Planning and Exploitation Company), or SAED, which was accompanied by a policy on settling families from the Middle River Valley. To compensate for the inefficiency of the former primary irrigation perimeters implemented, secondary and tertiary perimeters of these large areas were put into place in the 1970s. Water level management then became possible at the plot level and made it possible to record a yield of about 4–5 tons of paddy per hectare over a rice cycle.

At the same time, in order to extend the irrigated areas at more reasonable costs, the SAED facilitated the establishment of Village Irrigated Perimeters (VIP) in the Upper Delta: financing the acquisition of floating motor pumps by village groups, and support for leveling work. The investment costs of these small perimeters were five to eight times lower per hectare than in the large perimeters for equivalent paddy yields, which at the time contributed to Upper Delta family food security. However, the frequent absence of a drainage system condemned some of them to abandonment after only a few years (Dahou, 2009; author surveys).

The construction of two major works on the Senegal River was carried out during the 1980s. The downstream Diama dam aimed at preventing saltwater intrusion in the dry season (during the low-water period) was completed in 1986. The upstream Manantali reservoir dam was completed in 1988, making it possible to control the river's flooding and ensure the availability of water all year round for irrigated rice and vegetable cultivation. In parallel with these two dams, additional work was undertaken in the 1980s in the Lower Delta, along the Lampsar, a branch of the Senegal River. This concerned both the primary network, with the separation of irrigation and drainage networks and the installation of electrified pumping stations in the primary canals and the main water evacuation ditches, and tertiary perimeters, with the establishment of a dense dike network and leveling of all the fields. All of this hydraulic infrastructure allowed for precise flood regulation and stopped sea water from rising into irrigated areas. Far from the ambitious objectives initially set in terms of surface area, nearly 14,000 hectares of land in the delta were nevertheless under total water control by the end of the 1980s (Seck, 1991; Maiga, 1995).

5.3.2   Disruption of the Delta's agriculture, a determining factor in rural adaptation to drought

The installation of these dams and the total control of the river's flooding had a spectacular effect on the increase of area dedicated to irrigated rice cultivation during the 1990s (according to SAED data, from 14,000 ha at the end of the 1980s to 38,000 ha at the end of the 1990s). This was especially true in the Upper Delta, where there were many depressions well suited to irrigated rice cultivation (Fig. 5.5). In the Lower Delta, new rice-growing areas were planned, especially along the Lampsar River, with the SAED allocating irrigable fields of 0.2–0.3 ha per farmer. The maintenance of a minimum water level throughout the year and the dissemination of four-month rice varieties made it possible to grow off-season rice from March to July and henceforth made it possible to grow two rice seasons per crop year.

Off-season vegetable crops (onions, tomatoes) were also grown in the rotation between 2 years dedicated to rice. Their development was facilitated by the creation of SOCAS, *Société de Conserverie du Sénégal* (Senegal National Canned Foods Society), which, in the 1970s, set up its own irrigated area (550 ha) and a processing plant in Savoigne (Lower Delta) in order to meet Senegalese demand for tomato concentrate and to try to reduce imports. Due to the proximity of the SOCAS factory, which established contracts with producers, the cultivation of off-season tomatoes increased significantly in the Lower Delta. The extension of the irrigation and drainage network was also accompanied by an increase in the installation of small manually irrigated vegetable fields along the canals and drainage axes. With the cessation of rainfed crops imposed by rainfall deficits and the decline and disappearance of floodplain crops due to dam construction, these fields were partly devoted to certain crops that were previously grown under rainfed conditions (cowpeas, peanuts).

It seems that the development of flooded rice cultivation and irrigated crops (rice and vegetables) in the Upper and Lower Delta played

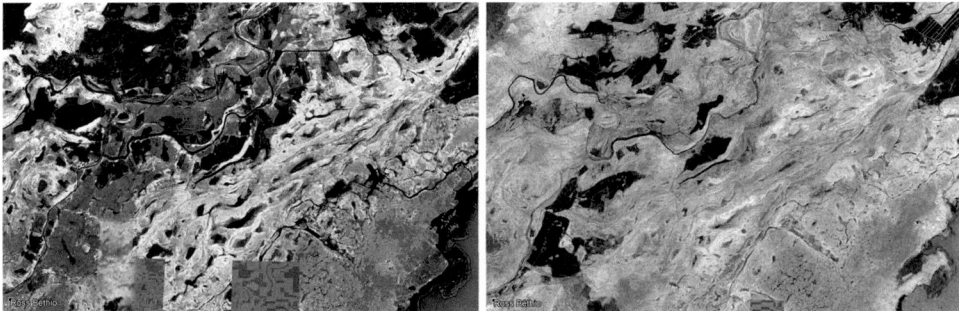

Fig. 5.5. Effects of dam implementation along the Senegal River on the extension of irrigated areas in the Upper Delta Left: December 1984; Right: December 1999. (From: Google Earth.)

an important role in agricultural production systems' adaptation to the severe drought of the 1970s and 1980s, at the cost of significant labor intensification, by offering the possibility of continued vegetable cultivation, growing another cereal (rice instead of millet), and having crops to sell (tomatoes in particular), which allowed for additional rice purchases. In addition to working for SOCAS, many families in the delta looked for additional income by trying to find work in Saint-Louis or continuing their artisan or commercial activities.

These transformations in delta agriculture, a real change in the state of this agrarian system to limit the effects of the crisis and climatic hazards, were based on systematic intervention by public bodies, in contrast to the situation that prevailed until the 1950s: (i) land management with the classification of delta land as "pioneer zones" as early as 1964, giving the SAED total control over land; (ii) organization of markets (monopoly on the purchase and sale of national cereals by the *Caisse de Péréquation et de Stabilisation des Prix* [Price Equalization and Stabilization Fund]), or CPSP; (iii) subsidizing inputs, access to credit, investment management, and agricultural planning; and (iv) management of irrigation water and even crop cultivation within the irrigated perimeters, according to a model based on the use of moto-mechanized machinery and chemical inputs, which are based on highly standardized technical itineraries, going as far as the SAED setting harvest dates. The result was producers relying heavily, both technically and financially, on the SAED (Seck, 1986) find themselves reduced to the role of mere executors within irrigated perimeters in which they would never have had the means to invest on their own.

On the other hand, the expansion of irrigated areas resulted in a reduction in areas dedicated to flood recession crops and dry-season grazing and led many herders in the Upper Delta to devote more time to crops and to settle down. A new relationship between crops and livestock was established, where the consumption of rice straw and bran by cattle in the dry season attempted to compensate for the decreased rangelands previously available in the floodplains. Santoir (1994) argues that the long-standing presence of hydro-agricultural perimeters offering fodder by-products and other sources of agricultural income may have contributed to cushioning the

massive decline and strong fluctuations in spontaneous fodder available to delta producers during the drought of the 1970s and 1980s, while stressing that transhumance, previously oriented toward the north and the major riverbed, was redirected toward the south, which was more watered. This transhumance also extended to the extreme southeast of the country, in the Terres Neuves region, where agricultural colonization had been underway since the 1970s (Devillers and Frissard, 2016).

5.3.3 A break in agricultural and land policy at the turn of the 1980s

The great level of continuity in the irrigation development objective (total water control at the plot level, lengthened duration of access to irrigation during the year, extension of irrigable surfaces) contrasts with the upheaval in production conditions and access to resources that producers have experienced since the 1980s.

During the 1960s and 1970s, the producers' technical and economic security was characterized by their close supervision by state bodies, first and foremost the SAED, both technically and financially, as well as in terms of access to irrigable land and attempts to regulate the domestic rice market in the 1970s. Created in 1971, ONCAD (National Office of Cooperation and Development Assistance) federated and administered all support functions for producers and aimed to promote national production. It was responsible for importing and storing imported rice, supplying and distributing subsidized inputs (up to 50%), and granting credit. From 1973 onward, ONCAD relied on the CPSP, which had a monopoly on the purchase and sale of national cereals as well as on cereal imports (including American food aid), and a regulatory system (wheat quotas, taxes) that enabled control of domestic cereal prices. The State tried to support the price of whole-grain rice offered to Senegalese producers, the form in which most local production is sold, by imposing an *ad valorem* tax on imported rice and through import quotas (Benz, 1996).

The Structural Adjustment Plan to which Senegal was subjected in 1981 and the resulting

New Agricultural Policy (NPA), initiated in 1984, disrupted delta farmers' productive framework from the 1980s onward, while hydro-agricultural planning continued in parallel. The bankruptcy of ONCAD in 1979 due to laxity in the reimbursement of cropping and equipment credits, which illustrates the rice sector's economic imbalance (Lavigne Delville, 1993), was accompanied by the SAED's gradual "disengagement" throughout the 1980s and 1990s and progressive delegation of water management to producers within the major irrigated areas, thanks to the formation of Hydraulic Unions, which became responsible for collecting water fees, maintaining canals and distribution works, and acquiring community equipment. The supply of inputs and the collection, processing, and marketing of rice were progressively transferred to private operators. Access to credit provided by the *Caisse Nationale de Crédit Agricole du Sénégal* (National Agriculture Credit Fund of Senegal), or CNCAS, had been based on short-term loans for inputs and medium-term loans for equipment, granted only to producers who owned land and were organized in Economic Interest Groups (EIGs), which may be family-based or linked to Hydraulic Unions. Producers also had to deal with a halt in input subsidies.

With the 1972 law on rural communities, which provides for the transfer of "pioneer zones" to "terroir zones," the SAED was no longer responsible for land management: the rural communities, via the intermediary of rural councils elected from among them, have become responsible for land allocation and management (Boutillier, 1989). While this decentralizing drive to transfer land management to the rural communities directly concerned may seem laudable, it has not been free of allocation inequalities, due to the clientelist drift linked to the new profitability that irrigation development potentially confers to land (Dahou, 2004).

5.3.4 New vulnerabilities despite irrigation progress: specialization, debt, and land insecurity

The extension of irrigable areas has been therefore above all based on the implementation of private perimeters, which experienced a rapid rise from the 1980s onward, particularly in the Upper Delta. While public irrigation perimeters hardly progressed (reaching less than 13,000 ha in the mid-1990s), private irrigation areas increased from less than 4000 ha to more than 29,000 ha in one decade (from 1987 to 1996) (Dahou, 2009), thus making up the better part of the extension of irrigated areas during the 1990s.

These private perimeters, which benefited from much more flexible management than the large areas of the 1970s, are of lower quality and less sustainable and often do not allow rice yields to be as high as those now achieved in the large perimeters. The increase in irrigated areas is also reflected in producers' increased specialization in rice cultivation, particularly in the Upper Delta, which makes them more vulnerable in the event of a poor harvest. Combined with rising production costs from the end of input subsidies and the fact that producers are now responsible for hydraulic unit maintenance and equipment costs, the deployment of private perimeters has been accompanied by a rapid increase in producers' debts. Faced with the enormous capital needs (in the short and medium term) involved in the extension of irrigated areas and credits being granted solely on the basis of land ownership, and having no prior expertise as to the technical viability and financial solidity of these borrowing structures, the CNCAS found itself deep in debt at the beginning of the 1990s (Dahou, 2009).

The devaluation of the CFA franc in 1994, which increased input prices, and, that same year, the dismantling of the CPSP, which had been buying local rice at a sustained price (through an *ad valorem* tax of 15% and import quotas), were also felt by rice producers in the Upper Delta. While an *ad valorem* import tax of 15 FCFA/kg was maintained, the relative evolution of rice and input prices proved very unfavorable to producers, who saw a 25% drop in the price of rice in real value between the early 1990s and the 1996–2007 period (Fig. 5.6).

In this context, the expansion of irrigated rice-growing areas between the mid-1980s and the mid-1990s seemed to have increased the country's rice autonomy, with national production covering on average a third of needs over this period. However, national production fell to less than a quarter of national consumption between 1995 and 2007, increasing by only 25%

over this period, while at the same time imports exploded, more than doubling over the same period (Fig. 5.7).

Our surveys also show that the new production context, access to land, and indebtedness of family producers provided favorable conditions to production systems with a high availability of capital and the necessary relationships and network of influence to allocate rights to large areas of farmland through rural community authorities,

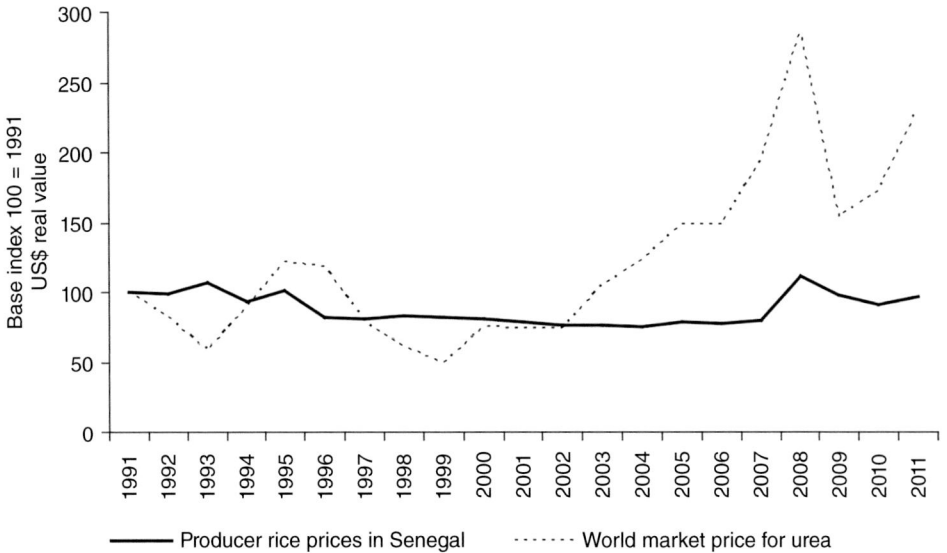

Fig. 5.6. Trends in producer rice prices in Senegal (1991–2011) Base 100 in 1991, in 2010 US$. (Data: FAOStat and World Bank; graph: authors.)

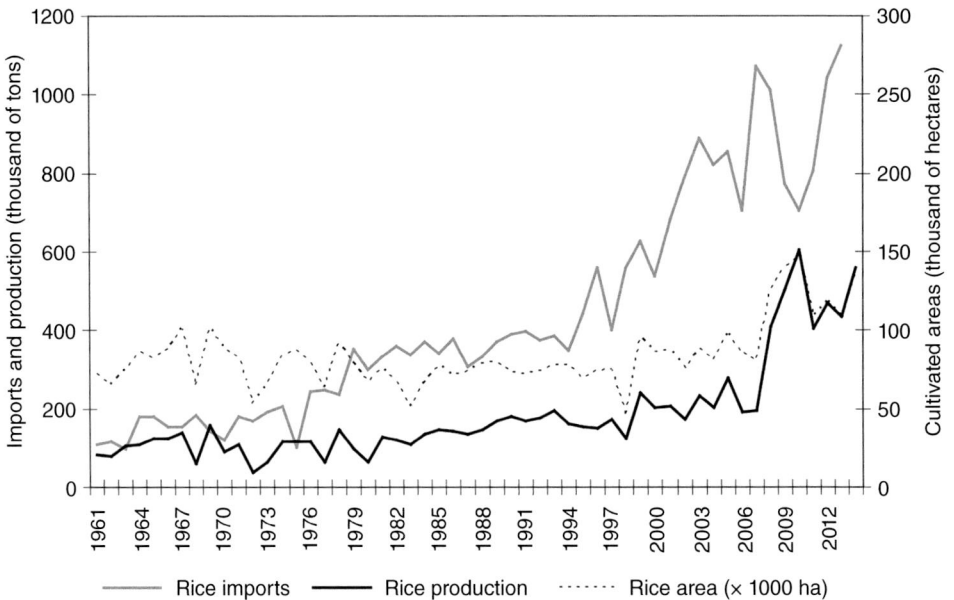

Fig. 5.7. Rice area, production, and imports in Senegal (1961–2014). (Data: FAOSTAT; graph: authors.)

resorting if necessary to clientelist practices, about which Boutillier (1989) already expressed fears nearly 30 years ago. If these practices were not sufficient, the capital owners made amicable arrangements with the numerous villagers who did not have the capital necessary to carry out secondary and tertiary irrigation development: the latter made requests for land from the rural community, which was then entirely developed by the investor, who only handed over part of it to them and reserved the exploitation of the main part of the irrigated fields for himself, using hired labor. From the 1990s onward, the construction of private perimeters in the Upper Delta has thus been accompanied by a first wave of developing patronal systems led by the wealthiest families, and even capitalist systems (echoing the observations of Boutillier, 1989) led by Senegalese and, to a lesser extent, international investors, and with it the widening of inequalities within delta agriculture.

5.4 From Adapting to Climate Change to Strengthening Food Self-Sufficiency: Which Development Model for Delta Agriculture?

5.4.1 A new wave of hydraulic planning, influenced by the 2008 cereal price hike

Since 2008, when cereal prices soared and the food crisis ensued, the Senegal River Delta has undergone a new wave of hydraulic planning, initiated in 2009 by the *Programme pour le Développement des Marchés Agricoles au Sénégal* (Program for the Development of Agricultural Markets in Senegal), or PDMAS. This particularly concerned the Lower Delta, with the reinforcement of the Lampsar's flow thanks to the construction of the Krankaye canal from the Gorom (connected to the Senegal River), which was accompanied by a 2500-ha extension of irrigable areas on the dune lands, spread over five perimeters installed on several villages' farming territory. This irrigation development resulted in the allocation of land plots ranging from 2 to 100 ha within these new areas, accompanied by investment and equipment subsidies of varying levels depending on the areas allocated. This program included a partial retrocession of land to expropriated villages and the

allocation of fields of more than 10 ha to private operators with the resources to develop irrigation in larger areas. The initial land distribution objectives in these areas were approximately 50% for farmers organized in EIGs (with priority given to women and young people), 25% for small and medium-sized enterprises (10 or 20 ha fields), and 25% for agro-industry (from 20 to 100 ha) (PDMAS, 2009).

The Millennium Challenge Account (MCA) conducted from 2010 to 2015 (US government funding, $540 million) aimed to complete and strengthen the delta's hydraulic infrastructure by:

- securing the 40,000 ha that could already be irrigated with adductors;
- extending the irrigated area by 10,000 ha; and
- transforming one of the Senegal River's emissaries, the Djeuss, used until then for irrigation, fishing, and breeding (watering, grazing in the dry season on land invaded by flooding in the rainy season) into the main drain canal for the irrigated areas of the Upper and Lower Delta.

In the Upper Delta, all the rice-growing areas' drainage networks were connected to the main drain (Djeuss) and this work seems to have made it possible to secure gravity irrigation on the fields furthest from the main canals and make small private areas, which had been summarily planned in the 1990s and abandoned after a few years, suitable for cultivation, thus contributing to the extension of the cultivated irrigable areas. In the Lower Delta, the damming up and change in use of the Djeuss upset the hydraulic scheme and was accompanied by the establishment of two "compensatory" canals. These not only serve the fields previously irrigated by the users from the Djeuss but also offer the possibility of extending irrigation to new areas, in depressions (basins) and above all on the levels of old marine terraces and dune lands, provided that one is equipped with a motor pump unit.

The reinforcement of irrigation capacity in the delta is part of the Senegalese government's desire over the past 10 years to strengthen the country's food self-sufficiency, particularly by increasing rice production. This has been the case since 2008, with the *Grande Offensive Agricole pour la Nourriture et l'Abondance* (Great Agricultural Offensive for Food and Abundance), or GOANA, and the *Programme National d'Autosuffisance en Riz* (National Rice Self-sufficiency Program), or PNAR, based on granting subsidies for equipment (donations of motor pumps) and inputs (fertilizer subsidies), and the cancellation of producers' debts in order to allow them to take out new loans with CNCAS. In parallel to hydraulic planning, these policy measures have been extended since 2014 with the implementation of the *Programme d'Accélération de la Cadence de l'Agriculture Sénégalaise* (Senegalese Agricultural Acceleration Program), or PRACAS, which aims, among other goals, to achieve Senegal's self-sufficiency in rice and onions by the end of 2017, with an envisaged increase in national paddy production of 1.6 million tons (MAER, 2014), equivalent to the whole country's rice imports and twice as high as national production in 2014 (Fig. 5.7). This program includes the distribution of subsidized equipment, input subsidies, cancellation of producers' debts (budget of nearly 20 million euros), and support for the planning of new perimeters on 5000 ha over 3 years (budget of 7.5 million euros). In 2015, 2000 ha were newly developed with the support of the SAED, 50% for family EIGs, and 50% for investors; in 2016, 1000 ha were developed and allocated in lots of 5, 10, 20, and up to 100 ha.

5.4.2 Less and less regulated producer prices and unequal access to credit

Since 2000, Senegal has aligned the tax on imported rice with the Common External Tariff of the West African Economic and Monetary Union (WAEMU) (10%) and abolished the import quota system. The price of rice for Senegalese producers is therefore directly indexed to the price on the international market. After about 15 years of decrease, the price for Senegalese producers has recovered, in real value, to its level in the early 1990s (Fig. 5.6), thanks to the increase in world rice prices from the late 2000s on. This price context probably also helps explain investor interest in the rice sector. Rice marketing is now led by the Senegalese private sector. Initiatives are beginning to emerge to enable family farmers to better control the sale price and marketing of their rice, such as a warehouse recently set up in the northern part of the Lower Delta,

supported by Swiss cooperation (Fig. 5.8), which hulls, bags, stocks, and will soon be grading (grain and broken rice) and sorting rice according to variety and quality. These initiatives aim to adapt to the demand of Senegalese urban markets, where rice represents more than half of the cereals consumed (Fall, 2010).

Senegalese onions still benefit from an import tax of 26.5%, which helps support the producer price, but they face strong competition from imports of the end of stocks from Dutch producers sold at very low prices on the international market (David-Benz *et al.*, 2010). Tomato production, which is mainly intended for industrial processing (double-concentrated), is sold under contract to SOCAS (now private) with a producer price set within the framework of the *Comité National de Concertation de la Tomate Industrielle* (National Committee of the Industrial Tomato Sector) (Fall *et al.*, 2010). The market for fresh tomatoes and other vegetable products is not regulated and is subject to significant price fluctuations and direct competition from other Senegalese vegetable production areas, primarily the Niayes region.

Fig. 5.8. Hulling, bagging, and storing rice in a village collective warehouse. (Photo: authors, 2016.)

Rice producers can obtain inputs from private suppliers by taking advantage of a harvest advance (CNCAS vouchers) in the form of credits centralized at the EIG level and then at the Hydraulic Union. In theory, repayment is made just after the rice harvest, but in practice, it can be later: since credits for the following season are only granted on condition that all producers belonging to the same EIG have all repaid their credit, delays in repayment can sometimes lead to a postponement to the following season. According to our surveys, the heterogeneity of economic situations from one producer to another within the same EIG leads to amicable exchanges between producers: some do not use all of the inputs requested (aligned with a management sequence per hectare determined by the SAED) in order to give them away (resale or in exchange for the use of a motor pump, for example) to better-off producers who can use these fertilizers on some of their fields located outside the EIG's boundaries. A similar credit system is set up by SOCAS for tomato cultivation.

Access to credit from CNCAS for the purchase of inputs is much more hypothetical outside the rice-growing area, especially for vegetable cultivation on the dunes, and leads farmers to adopt other strategies: vegetable production that is less demanding in terms of cash flow, self-financing thanks to (i) external activities or by playing on the cash flow complementarities between rainy season and off-season production facilitated by the staggering of harvests between different agroecological areas, (ii) renting part of the land for the rainy season or off-season cropping, (iii) using livestock (for those who have it) as a source of cash, and (iv) using microcredit but at unsubsidized rates of 10–14%.

5.4.3   Adaptation, but reinforcement of inequalities within family farming

Since the 1990s, family farming in the Upper Delta has largely specialized in rice cultivation in the gravity-fed irrigated fields of the large perimeters (Fig. 5.9d; photo 5.1), some of which were developed in the 1960s, and most of which were transferred to the Hydraulic Unions in the 1980s. According to our surveys, each nuclear household now generally has 2–3 ha in these large paddy fields, where two rice crops can be carried out per year, supplemented by 0.5–1 ha in the small paddy fields (VIP), where they practice one rice crop per year and often some vegetable crops (Fig. 5.9b,c). The most precarious young households have often inherited only 0.5 ha in a large perimeter dedicated to rice production, and practice vegetable crops to a greater extent (on 0.5 ha) by renting a motor pump and then doing furrow irrigation (Fig. 5.9a). With the exception of Fulani families, which are often organized into extended families where the collective herd of 30–40 cattle may be entrusted to one among the younger siblings, very few own cows.

Also, since the 1990s, family farms have coexisted with systems implemented by farmers sufficiently well-off to have applied for land allocations at the time and gradually developed private areas, sometimes of several dozen hectares, by ceding the use of part of the land to investors (wealthy urban dwellers from Dakar, Rosso, or Richard Toll) in exchange for gradually developing irrigation on a few hectares. In this way, they complete 4–5 ha, two-thirds of which are located in large, irrigated perimeters and one-third in small ones, with a private area dedicated to rice cultivation (one season, often off-season) that can reach up to 20 ha and is equipped with a motor pump.

In the Lower Delta, our surveys show that the most precarious households (often young people) have less than 1 ha, a motor pump in joint ownership, and no animals. Eighty percent of their farmland is located in the large irrigated perimeters, where farmers alternate 1 year of two rice seasons with 1 year of onion cultivation [rice/rice//onion]. The remaining area is rented on dune land dedicated to off-season vegetable crops that require little capital (eggplant, okra, cucumber, etc.), and whose earlier harvest plays a key role in their cash flow calendar (Production System A, PS A). Slightly older households, equipped with a motor pump and backpack sprayer but still without livestock, are only marginally better off. They farm their own 1–1.2 ha (Production System B, PS B), with less than half of their area devoted to the same vegetable crops, which are riskier and more costly than rice. Rice therefore occupies a comparatively important place in their cropping choices. Agricultural income alone is not enough to support

Fig. 5.9. Irrigated vegetable and rice crops in the Upper Delta. Small areas (a), private areas (b and c), and large areas (d). (Photos by authors, 2016.)

these families, who have to turn to complementary salaried activities to generate a total income per worker that is a little higher than that of the region's salaried farmers (Fig. 5.10).

Other family farms are not limited in land (0.45–0.6 ha per family member compared to 0.3 ha in the previous category) but are mostly located on the dunes and rarely in the depressions. They are equipped with a motor pump, one or two backpack sprayers, and a herd of 10–20 goats. All these conditions have led them to specialize largely, if not exclusively, in vegetable crops, which occupy at least three-quarters of their cropping area (PS C and PS D). They combine crops that demand cash flow (onions, chilis) and organic manure (chilis) with less risky ones (African eggplant). They nevertheless rent out half of their land because they do not have enough cash to farm it themselves.

Some family farmers with more land (0.8 ha per family worker, PS E), in balanced proportions between rice fields ([rice/rice// onion] or [rice/rice//fallow land for 1 year]) and dune land dedicated to vegetable crops, have progressively evolved toward patronal systems (more than half of the agricultural labor is provided by salaried employees). Rice occupies about a third of their land and the remaining two-thirds are devoted to more costly crops (tomato, onion, chili). They are also able to do some sheep fattening.

The most affluent families, the only ones to have kept cattle and goat herds to be run in the extended family, plus sheep for fattening, have had the necessary support to be allocated large areas (5–10 ha per family member), most of which they rent out for the time being, which provides them comfortable land rent (PS F). This

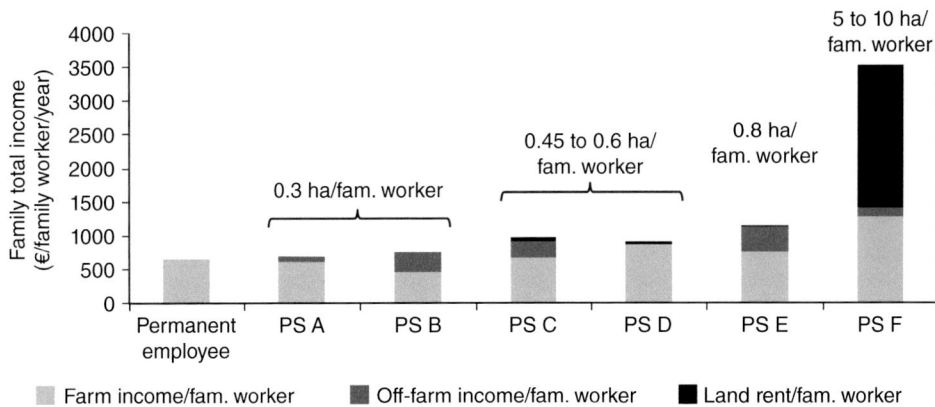

Fig. 5.10. Level and composition of income per worker in the main production systems. (Data: surveys, 2016.)

land and financial wealth have enabled them to specialize to a large extent in vegetable crops (onions, chilis).

The differences in income per worker between the poorest and the best-off families (taking into account all sources of agricultural and off-farm income and land rent) can vary by a factor of one to five (one to two between the family system and the patronal system without land rent, Fig. 5.10).

Our surveys show that off-farm income (salaried employment) is not negligible in Lower Delta families' total income and plays a key role in financing rice and especially vegetable crops. The centrality of salaried labor in production system functioning can be seen in Fig. 5.11. The vast majority of farms depend on it, either on a daily basis to cover certain labor peaks (harvesting in particular) or through employing workers either seasonally (for the off-season, particularly for vegetable crops) or on a more permanent basis (employees present 10 months out of 12 on the farm). While the daily workforce is recruited locally (young workers, often from poorer households), the seasonal or permanent workforce is most often from other regions of Senegal, mainly from Casamance, and is made up of young workers who migrate to the Senegal River Delta region during the dry season when the agricultural workload is lower in their region of origin. Patronal systems (e.g. PS E) are those that mobilize the most salaried labor.

5.4.4 Standing by the choice of a multi-speed agriculture for the delta

In the Upper Delta, the strong progression of private perimeters since the 1990s has resulted in the development of capitalist production systems (farms with employees) specialized in rice production: from 40–50 ha with two crops per year, up to over 100 ha dedicated to intercropped rice cultivation, with cattle breeding of several dozen heads (capital accumulation in the form of livestock) entrusted to a herdsman and complemented by service provisions (tilling, harvesting) to make their large-capacity motorized equipment (tractors, tillers, combine harvesters) profitable.

In the Lower Delta, the recent irrigation development and public policies have resulted in a new wave of competition for land, in which the owner-farmers and investors are particularly advantaged by their high availability of capital (loans or self-financing), equipment, and seasonal cash flows (inputs, labor, etc.). The Lower Delta has thus experienced a wave of development of capitalist systems over the past few years, thanks to the strengthening and extension of large-scale hydraulic infrastructure, but also the land allocation methods and investment support from public authorities (through the SAED and rural communities).

The extension of the primary network and reinforcement of its flow has led to a surge in land allocation requests and, it seems, to most of the land in the Lower Delta being allocated. This particularly concerns the dune areas, part of which have already begun to be exploited, but also the depressions that were not, until now, served by the primary canals.

According to our surveys, capitalist forms of agriculture materialize in different types of farms, mostly owned by domestic investors:

- Retired agricultural officials from the Lower Delta delegate the management of less than 10 ha of vegetable crops to salaried employees (see below, PS G).
- Investors from Dakar, Thiès, or Touba, who often have other farms in other regions of Senegal, gradually develop irrigation over 10–20 ha located on the dunes with sprinkler or drip systems for specialized vegetable crops

intended for national niche markets with high value added (potatoes, red cowpeas, etc.).

- These same investors, or foreigners posted in Dakar, may also have had access to land in the depressions (basins) over larger areas (25–30 ha), which are currently used for rice at a lower cost (see below, PS H) in order to be able to, in the long term, combine (after sufficient desalination of the land) rice cultivation and vegetable crops.
- Some investors from the region (including local elected officials) have been able to benefit from vast allocations in several fields of 30–70 ha, either in depressions or on the dunes where they can be equipped with center-pivots (Fig. 5.12), and combine

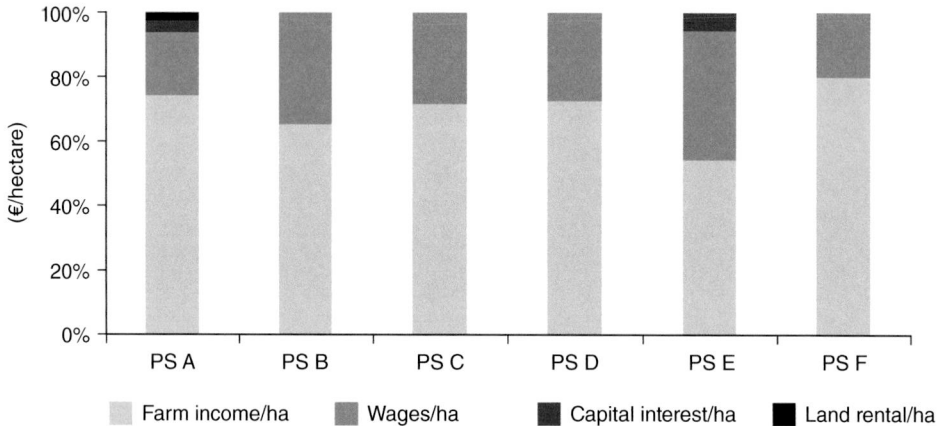

Fig. 5.11. Distribution of the value added created per hectare between farm income, farm worker wages, land rental, and interest on borrowed capital. (Data: surveys, 2016.)

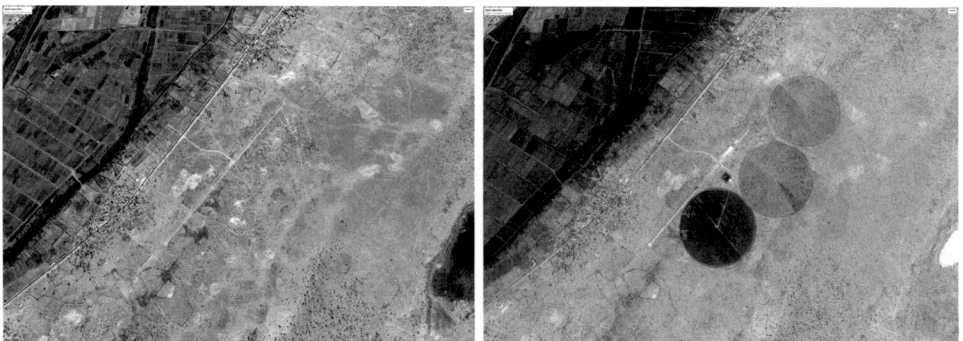

Fig. 5.12. Establishment of a capitalist farm on dune land in the southern Lower Delta in 2015. Equipped with three center-pivots, combining vegetable and poultry production (left: February 2015; right: July 2016). (From: Google Earth.)

rice cultivation, vegetable production, and sometimes confined livestock (broilers) or feedlots (beef cattle), maximizing by-products and possible complementary fodder production (corn, pulses).

Investment is most often limited to possible land clearing, fences, grading, and installation of small dikes, as well as irrigation equipment (motor pump unit, pipes, center-pivots, sprinklers, drip system, etc.), and much more rarely motorized equipment (tractor, offset for leveling, sprayer, combine harvester, etc.). To do this, they call on service provider companies that are developing in the region while relying on salaried labor for many tasks that remain manual.

Calculations based on our surveys of investors show that the purchase and exploitation of 30 ha for irrigated rice production (Production System H or PS H) is profitable (financial evaluation conducted over 20 years) for real interest rates of less than 17%. However, real bank interest rates in Senegal have not exceeded 8% in the last 10 years (World Bank). For a real interest rate of 5%, the cumulative profit over 20 years (after updating) is 40,000 euros, which is nevertheless sensitive to rice price variations. The high internal rate of return of this type of investment project is a testimony to its low-risk character, which leaves investors with the latitude, once the land has been desalinated by several years of irrigated rice cultivation, to devote part of the surface area to more capital-intensive production, which is riskier and whose profitability is more sensitive to interest rate variations, but potentially much more remunerative for moderate interest rates.

A final category (5) includes agricultural enterprises owned by international investors, most often in place since the 2000s, but whose exploited areas have been extended in recent years. Now in charge of several hundred hectares on the dunes, divided into different blocks (with two-year sub-leases for plots beyond 100 ha; Photo 5.2), they own their own motorized equipment (Fig. 5.13) and are equipped with center-pivots, sprinklers, or drip irrigation. These producers are specialized in vegetable or fruit production for export (melon, watermelon, squash, bananas, citrus fruits, etc.), whose off-grade products are sold on the local or national market.

Our surveys and the information collected by S. Kamara (2013) provide a (non-exhaustive) overview of the foreign-owned agricultural companies currently operating in the Lower Delta (Table 5.1).

Favored by their self-financing capacity for tertiary hydraulic planning and equipment purchase, these forms of capitalist agriculture do not, however, appear to be more wealth-creating than family farming for equivalent production types. According to our surveys and calculations, for a given production system, a field located in the clay basins cultivated with two rice seasons per year (rice/rice) generates a gross value added per hectare that is 25% lower than a biennial succession (rice/rice//onion) and at least 35% lower than an annual succession comprising two vegetable crop seasons (e.g. tomato/watermelon + squash) practiced on dune lands (for which, however, the use of a motor pump is necessary). The net values added recorded per hectare at the production system scale are therefore all the higher as the proportion of land used in the basins (devoted at least 1 year out of 2 to rice cultivation) is reduced, and the production system is specialized in vegetable production (with a direct correlation to families' capital and cash flow). For the same degree of specialization (e.g. 100% of the land devoted to vegetable production), it appears that the creation of wealth per surface area unit is more or less the same for similar products (tomatoes, onions, chilis), between a capitalist type of production system with a high availability of capital (PS G) and a family production system (PS C). Strict rice specialization, which can only be found in the Lower Delta within capitalist production systems (PS H), creates much less wealth per unit area (Fig. 5.14).

5.4.5 The deployment of new project formats with Public-Private Partnerships (PPP): guaranteeing greater equity?

Two main Public-Private Partnership (PPP) projects are underway in the Senegal River Delta, which are financed by foreign donors and plan to address different producer categories by proposing areas to be exploited that vary from a few hectares to lots of 50 or 100 ha. Fully financed by donors for primary planning, these projects provide for variable levels of subsidies from one

Fig. 5.13. Pumping station, high-capacity motorized equipment, and minibus transporting farm workers from a foreign-owned farm in the Lower Delta. (Photos by the authors, 2016.)

producer category to another for secondary and tertiary planning, depending on the surface area of the lot allocated. One of these projects is conducted within the PDMAS framework and concerns the extension of 2500 ha of vegetable crop areas in the Lower Delta (Photo 5.3), while the second (the 3PRD: *Projet de Promotion des Partenariats Rizicoles dans le Delta*, Delta rice partnership promotion program) concerns the planning of an irrigated perimeter of 2500 ha in the Upper Delta (Fig. 5.15).

Within the PDMAS framework, the allocations concern 2500 planned hectares, divided into five perimeters installed on the land of five villages in the Lower Delta, of which only half is reserved for family EIGs. Our investigations within the five areas show that field allocating methods differ from one village to another, ranging from allocating fields of less than 10 ha to the only families in the village that had a right to use the land before development (as in the case of the Polo area), to allocating two-thirds of the planned areas to a foreign contractor (the SCL in the case of the Massar Gabou area). Despite the fact that the project covers most of the infrastructure costs (a subsidy of 80% of the costs) and equipping family farmers (provided they use drip or micro-sprinklers), the capital needed to cover planning costs remains high (around 1200 euros per hectare, covering 20% of the total costs), equivalent to all the income earned by a family farmer for 1 or even 2 years (Fig. 5.11). Developing irrigation in fields of more than 1 ha thus remains out of reach for most family farmers from these villages, especially those who do not have their own motor pump, family labor, or cash to hire the necessary paid labor. When poorer households manage to form a collective

EIG to do a group application, the areas allocated do not exceed 0.1 ha per family, and some of these EIGs have not yet been able to begin a crop year due to lack of equipment. For lack of a better option, they hope to rent out their land.

Still in planning in 2016 (Fig. 5.15), the irrigated perimeter installed in the Upper Delta under the 3PRD initially covered 2500 ha,

Table 5.1. Characteristics of some of the foreign-owned agricultural companies established in the 2000s in the Lower Delta. Data: authors' surveys 2016 and Fert and Radzik (2016); Kamara, 2013.

Company	Main source of capital	Date established	Available agricultural area (assigned+ rented) *(year)*	Cultivated agricultural area *(year)*	Main productions	Main product destinations
GDS	France	2003	?	200 ha *(2011)*	Tomatoes, cherries, asparagus, sweet corn	European Union
Soldive	France	2006	H165 ha leased *(2016)*	> 120 ha *(2016)*	Charentais melon	France
SCL	France, Morocco, England	2006	300 ha assigned + 200 ha rented *(2011)*	500 ha *(2011)*	Sweet corn, asparagus, squash, peanuts	England
STS	Italy	2007	200 ha assigned *(2011)*	110 ha *(2011)*	Tomatoes	Italy

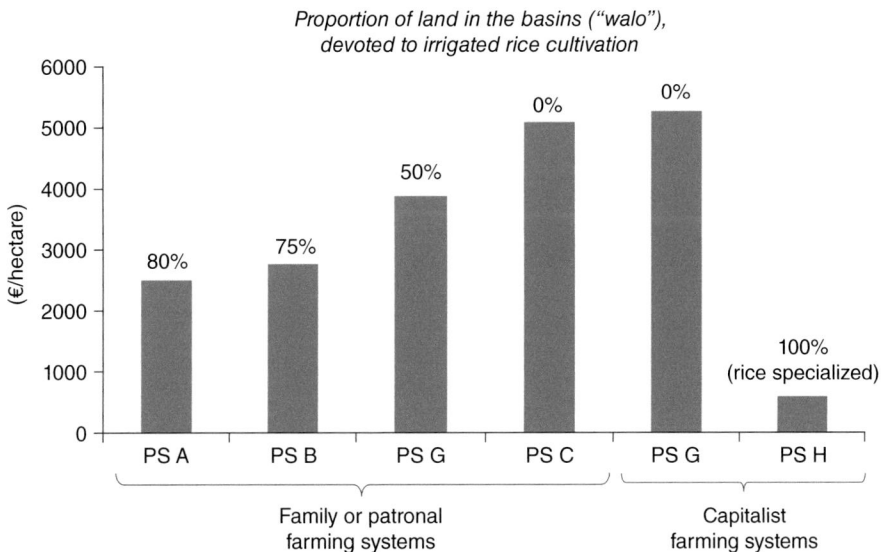

Fig. 5.14. Net value added per hectare for different Lower Delta production systems, characterized by the varying importance of rice production. (Data: surveys, 2016.)

Fig. 15.15. Crops and works in the irrigated areas of the Delta. Chili cultivation (patronal system) in one of the PDMAS perimeters in the Lower Delta (left) Work in progress within the future 3PRD perimeter in the Upper Delta (right). (Photos by the authors, 2016.)

which was reduced to 2150 ha to stay within the initial budget. It provides for the allocation of plots ranging from 5 ha to 100 ha, the distribution of which was based, in 2016, on the following breakdown: 51 allottees would have 5 ha plots (255 ha, 14% of the total area), 36 allottees would have 10 ha plots (360 ha, 20%), 32 would have 25 ha plots (800 ha, 45%), five 50 ha plots (250 ha, 14%) and one 100 ha plot (6%). Here, too, the investment subsidy rate, which decreases as the allocated area increases (95% for 5 ha, 85% for 10 ha, 70% for 25 ha, 40% for 50 ha, and 0% for 100 ha), is intended to provide greater support to the allottees of the "smallest" lots. For a 5-ha plot, the initial investment nevertheless comes out to more than 1000 euros, an amount inaccessible to the Upper Delta family producers who are limited by their surface area, even if they join together in an extended family EIG. The project's specifications also require future producers to produce a minimum of seven tons of paddy rice per hectare per year (or risk losing their right to farm) and pay a fee of 150 euros per hectare per year. The creation of the land reserve necessary for the establishment of this perimeter was done by expropriating the villages that had previously held a right of use over these fields, with compensation based on production history and equivalent to the value of 2 years of cultivation. Farmers who had not cultivated these areas for more than 2 years received no compensation.

Ironically, the PDMAS is in fact inspired by the example of the village of Thilène in the northeast of the Lower Delta. Irrigation development in this village is an emblematic illustration of strategies underpinned by a very different rationale when compared to the mode of allocation of land plots provided for in the PDMAS and particularly the 3PRD. In Thilène, farmers have organized to collectively access the capital necessary for the development of irrigation in dune lands, while at the same time controlling land allocations and guaranteeing greater equity between villagers in the allocations. Faced with the State's disengagement, the village association has been involved on several occasions since the end of the 1990s in promoting the extension of irrigable areas for the Thilène villagers with a great degree of autonomy.

The first initiatives, self-financed at the village level, concerned the creation of two new small areas (20 ha and 30 ha) intended primarily for young people. In the 2000s, the desire to develop a larger area (400 ha), which would not only benefit the wealthy families of Thilène, led the village association to collectively seek additional external financing from investors and proceed with a real dialogue on land allocation. These outside investors finance part of the irrigation infrastructure in exchange for the right to use the land for a limited period of time. The irrigable areas are then allocated to different families in the village, accounting for the areas they have transferred to the collective to make up the 400-ha perimeter; the capital, equipment, and work time they have made available during the construction of the infrastructure; and their status, which is considered to be a priority (young people, families with less irrigable land until that point).

5.4.6 Worrying environmental and health effects

Livestock farming has been in major decline in the Senegal River Delta for several decades. While the long period of drought from 1970 to 1990 led farmers and herders in the delta to reduce their herd size, the disruption brought about by hydraulic planning in the Upper and Lower Delta land use has greatly reduced the areas devoted to pastoral activities (gradual reduction of grazing areas paced by the progression of irrigated fields; fencing off of the growing areas occupied by the patronal and capitalist farms, making them inaccessible to herds even during intercropping periods [sometimes lasting several months]). Farmers are forced to rely more and more on their own fodder resources for their animals: collected crop residue (straw, leaves, etc.) and grassy fallow land. Family farmers who still have small herds (small ruminants and, more rarely, cattle) are therefore also those who have the largest areas of land and can afford not to cultivate their land continuously throughout the year, preferring a single season (rice or vegetable crops) followed by 4–6 months of fallow land. Keeping animals fed with crop by-products close to their homes (Fig. 5.16) allows some organic manure transfer (provided that a cart is available to transport the dry dung), with priority given to certain vegetable crops (chili pepper). Most fertility inputs are nevertheless based on synthetic fertilizers, with consequences for both the soil's organic matter content and families' dependence on purchased inputs.

The upheaval of the hydraulic scheme implemented within the framework of the MCA, although it contributes to extending surface area and securing water access for family farmers, is nevertheless accompanied by worrying environmental and health effects. The Djeuss is now an outlet for water laden with pesticide residues and chemical fertilizers, but it does not yet cen-

Fig. 5.16. Feeding the cattle kept close to houses with rice cultivation by-products. a: transport of rice straw; b: housing; c: storage of rice straw. (Photos by the authors, 2016.)

tralize all the Lower Delta drainage water, some of which is still directed toward natural depressions. Furthermore, the project has not effectively put an end to some of the productive activities carried out on the Djeuss before its transformation into a main drain. Fishing activities are still carried out (Fig. 5.17) with significant health risks, notably by Malian fishermen who set up camps along the Djeuss and then ship the dried fish to Mali. The drainage axes are also still used as an irrigation water source for small vegetable crop fields by farmers who do not have fields near the canals.

The extension of canals, drains, and irrigated areas also contributes to increasing the space for mosquito development (which the distribution of mosquito nets only partially addresses) and the risk of bilharzia (whose intermediate host is a freshwater mollusk), the first cases of which appeared after the dams' construction and are tending to multiply (Observatoire de l'Environnement, 2006). Finally, the decline in tree species observed since the 1970s does not seem to be solely attributable to drought: the now irrigated depressions, which used to be invaded by floods and were home to a dense population of *Acacia nilotica*, has contributed to the virtual disappearance of this species in the delta (Roquet, 2008).

5.5 Conclusion

Although astutely adapted to difficult, unpredictable rainfall and river flooding conditions, agriculture in the Senegal River Delta was deeply affected by the drought of the 1970s and 1980s. Without concealing the disruption caused to the delta's ecosystems by all the hydro-agricultural planning implemented in the region since the 1960s, this has, in this climatic context, played a central role in the maintenance and development of the region's agriculture and its adaptation to drought. However, irrigation is far from having benefited only family farming. The

Fig. 5.17. Fishing activities in the drains of the Delta water system. (Photos by the authors, 2016.)

State's disengagement from land management and investment support from the mid-1980s onward, favored the most capital-rich producers and contributed to the first wave of business development of capitalist systems in the Upper Delta from the 1990s onward, creating the conditions for sufficiently remunerative investments for the Senegalese urban elite. The strengthening of primary hydraulic infrastructure in the delta in the last 10 years has already led to a second wave of expansion of these forms of agriculture, which are now creating unprecedented land competition in the Lower Delta.

After decades of drastic deficits, the Senegal River Delta (like the whole northern part of the country) has seen a timid increase in rainfall since the early 2000s. Without returning to the 1946–1968 level, average rainfall at the Saint-Louis station over the 2001–2010 period was 311 mm (Ndong, 2015), which explains why families with the most precarious land have been trying for some years, and succeeding when rainfall exceeds 300 mm, to reintroduce rainfed crops on dune land. There is thus a clear competition in the Lower Delta for use of the dune lands, between, on the one hand, patronal and especially capitalist farms capable of making the necessary investments to connect to the irrigation and drainage network, and, on the other hand, family farming, which is concerned about the use and development of these areas on a multi-generational scale, relying on irrigation but also on livestock activities, and on the potential re-investment of rainfed crops if the rainfall rebound were to come true.

After the allocation of most of the Upper Delta land at the end of the 1990s, thanks to the new irrigation possibilities offered by the two dams built on the Senegal River in the 1980s, most of the land in the Lower Delta seems to have been in turn allocated in recent years, without villagers always being well-informed on the allocations decided by their rural council. Although a few Lower Delta villages have organized themselves to set aside some of their undeveloped land for future generations and avoid allocating it to wealthier families or investors, the land situation as a whole is particularly worrying.

Indeed, agricultural prices and most family farmers' labor productivity levels do not allow them to generate surpluses and accumulate capital quickly enough to invest in irrigation infrastructure and equipment, given the pace of the colossal primary hydraulic infrastructure implemented in recent years. They are thus totally disadvantaged in comparison to patronal farms and above all to the Senegalese urban elite (and even foreign investors), who are in a strong land allocation position thanks to their high investment capacity and clientelistic practices. Faced with this situation, the Public-Private Partnership-type projects that have been developed in recent years in favor of strengthening irrigation possibilities in the region, when they are not the result of village initiatives, are in fact only marginally aimed at the most disadvantaged farming families and would actually contribute to reinforcing social inequalities. These initial attempts should therefore be viewed with great caution, as other projects of the same nature are being planned for even larger areas, such as the *Projet pour le Développement Inclusif et Durable de l'Agrobusiness au Sénégal* (Senegal Sustainable and Inclusive Agribusiness Development Project), or PDIDAS, which would concern 10,000 ha on the banks of the neighboring Lake Guiers.

The "acceleration of the pace" in agriculture, a motto that the Senegalese authorities seem to have coined in the name of a legitimate objective, food self-sufficiency, appears to indeed be disconnected from the realities of family farming in the delta and a source of major land grabbing. In this context, while introducing the recent PRACAS program, the voluntarist words of Macky Sall, President of the Republic of Senegal, take on a particular resonance: "*If everyone gets involved, we will soon radically change the face of Senegal, thanks to agriculture*" (MAER, 2014). By wanting to go (too) fast, is there not a risk of excluding the majority of delta producers from these development processes and making the next generation of family producers insecure, contrary to the sustainability imperatives set out in the PRACAS?

References

ANSD (2008) *Enquête sur la Sécurité Alimentaire en Milieu Urbain 2008*. Agence Nationale de la Statistique et de la démographie. Dakar, Sénégal.

Benz, H. (1996) *Local and imported rice in Africa: the determinants of competitiveness. Les filières riz du Sénégal et de la Guinée face aux exportations de la Thaïlande et des États-Unis*, tome 2. École des Hautes Études en Sciences Sociales, Paris, France, 325 pp.

Boutillier, J.-L. (1989). Irrigation et problématique foncière dans la vallée du Sénégal. *Cahiers des Sciences Humaines* 25(4), 469–488.

Dahou, T. (2004) *Entre parenté et politique: Développement et clientélisme dans le Delta du Sénégal.* Karthala-Enda Graf Sahel, Paris, France, 364 pp.

Dahou, T. (2009) L'autosuffisance rizicole, chronique d'un échec annoncé. In: Dahou, T. (ed.) *Libéralisation et politique agricole au Sénégal*. Crepos-Karthala-Enda/Graf/Diapol, Paris, France, pp. 149–171.

David-Benz, H., Diop, M., Fall, C. and Wade, I. (2010) Onions: a booming production to meet urban demand. In: Duteurtre, G., Faye, M.B. and Dièye, P.N. (eds) *L'agriculture sénégalaise à l'épreuve du marché*. ISRA-Karthala, Paris, France, pp. 171–196.

Devillers, J. and Frissard, C. (2016). Évolutions du système agraire des Terres-Neuves de Koumpentoum, Sénégal, de 1950 à nos jours et perspectives d'évolution. Dissertation, UFR Agriculture Comparée et Développement Agricole, AgroParisTech/AFD, Paris, France, 142 pp.

Fall, A.A. (2010). Rice: disengagement of the State, soaring imports and . . . return of the State? In: Duteurtre, G., Faye, M.B. and Dièye, P.N. (eds) *L'agriculture sénégalaise à l'épreuve du marché*. ISRA-Karthala, Paris, France, pp. 57–81.

Fall, A.A., David-Benz, H. and Huat, J. (2010). Local tomato and concentrate production: the strength of contracts between farmers and industry. In: Duteurtre, G., Faye, M.B. and Dièye, P.N. (eds) *L'agriculture sénégalaise à l'épreuve du marché*. ISRA-Karthala, Paris, France, pp. 197–215.

Fert, M. and Radzik, L. (2016) *Analyse-diagnostic d'une petite région agricole dans le sud-ouest du Delta du fleuve Sénégal: une région caractérisée par un développement récent du maraîchage.* Dissertation, UFR Agriculture Comparée et Développement Agricole, AgroParisTech/AFD, Paris, France, 186 pp.

Kamara, S. (2013) *Développements hydrauliques et gestion d'un hydro-système largement anthropisé: le delta du fleuve Sénégal*. PhD thesis. Université d'Avignon et des Pays du Vaucluse/Université Gaston Berger de Saint-Louis, Avignon, France, 461 pp.

Lavigne Delville, P. (1993) The paradoxes of disengagement. In: Blanc-Pamard, C. (ed.) *Dynamique des systèmes agraires: politiques agricoles et initiatives locales: adversaires ou partenaires*. ORSTOM, Paris, France, pp. 217–238.

Le Borgne, J. (1988) *La pluviométrie au Sénégal et en Gambie*. Laboratoire de Climatologie, Université Cheikh Anta Diop, Dakar, Senegal, 28 pp.

Lericollais, A. (1975) Peuplement et migrations dans la vallée du Sénégal. *Cahiers ORSTOM Série Sciences Humaines* 12(2), 123–135.

Lericollais, A. (1976) Drought and the populations of the Senegal valley. In: Les Nouvelles Éditions Africaines (ed), *La désertification au sud du Sahara, Colloque de Nouakchott (17-19/12/1973)*. Les Nouvelles Éditions Africaines, Dakar-Abidjan, Senegal-Côte d'Ivoire, pp. 111–116.

MAER (2014) *Programme d'Accélération de la Cadence de l'Agriculture Sénégalaise (PRACAS), Volet agricole du Plan Sénégal Emergent (PSE)*. Ministry of Agriculture and Rural Equipment, Dakar, Senegal, 110 pp.

Maiga, M. (1995) *Le bassin du fleuve Sénégal: de la traite négrière au développement sous-régional auto-centré*. L'Harmattan, Paris, France.

Michel, P. and Sall, M. (1984) Dynamique des paysages et aménagement de la vallée alluviale du Sénégal. In: Blanc-Pamard, C., Bonnemaison, J., Boutrais, J., Lassailly-Jacob, V. and Lericollais, A. (eds) *Le développement rural en questions: paysages, espaces ruraux, systèmes agraires: Maghreb-Afrique noire-Mélanésie*. ORSTOM, Paris, France, pp. 89–109.

Ndong, J. (2015) Recent climatic evolution on the Senegalese coast. *Revue de Géographie de l'Université de Ouagadougou* 4(2), 151–168.

Observatoire de l'Environnement. (2006) *Annual Report on the State of the Environment and Natural Resources of the Senegal River Basin*. OMVS, Dakar, Senegal.

PDMAS (2009) *Mise en valeur de ha2500 dans la zone du Delta du Fleuve Sénégal: Présentation du Programme et du cahier des charges de l'agro-investisseur*. Ministry of Agriculture and Fisheries, Dakar, Senegal.

Poupon, H. (1976) Influence of the 1972-1973 drought on the vegetation of a Sahelian savanna of the northern Ferlo, Senegal. In: Les Nouvelles Éditions Africaines (ed), *La désertification au sud du Sahara, Colloque de Nouakchott (17-19/12/1973)*. Les Nouvelles Éditions Africaines, Dakar-Abidjan, Senegal-Côte, pp. 96–101.

Reboul, C. (1984) Dams against development? Les grands aménagements hydrauliques de la vallée du fleuve Sénégal. *Revue Tiers Monde* XXV(100), 749–760.

Roquet, D. (2008) Leaving to last: migration as a response to drought in Senegal? *Espaces, Populations, Sociétés* 2008(1), 37–53.

Santoir, C. (1994) Decadence and resistance of pastoralism. Les Peuls de la vallée du fleuve Sénégal. *Cahiers d'Études africaines* XXXIV(1–3), 231–263.

Seck, S. (1986) La maîtrise de l'eau et la restructuration sociale induite par l'organisation de la production irriguée dans le bassin du fleuve Sénégal. *Les Cahiers de la Recherche Développement* 12, 13–20.

Seck, S. (1991) Sur la dynamique de l'irrigation dans la vallée du fleuve. In: Crousse, B., Mathieu, P. and Seck, S. (eds) *La Vallée du fleuve Sénégal: évaluations et perspectives d'une décennie d'aménagements*. Karthala, Paris, France.

III

Adaptation and Resilience in Mountain Regions

6 Altitudinal Tiering, Diversity, and Irrigation: The Uluguru Mountains, Tanzania

Hubert Cochet[1]* and Thérèse Hartog[2]

[1]*Comparative Agriculture Training & Research Unit/UMR Prodig, AgroParisTech, Palaiseau, France; [2]ODEADOM, Paris, France*

Is the question of how to adapt agriculture in the South to climate change posed differently in mountain regions, where farmers must deal with altitude, slope, and the erosive phenomena that they facilitate, as well as tiering and sometimes communication difficulties? Throughout the world, such situations are too numerous, varied, and dissimilar for one to hope to answer this question in a global way. With steep hillsides entirely cultivated by a very dense agricultural population (250–350 inhabitants/km²), systematic planning of these hillsides, generalized irrigation, and diversity of production, the Uluguru Mountains—the only example of a mountain region dealt with in this book—are nevertheless an example rich in lessons[1].

6.1 The Uluguru Mountains: An Example of Densely Populated and Well-Watered Mountains

6.1.1. The Uluguru Mountains water tower

The Uluguru Mountains are a small, isolated mountain range on the edge of the eastern Tanzanian lowlands (Fig. 6.1), isolated from the historically less populated areas that surround it, but now fairly well connected to Morogoro—at least its western side (the Upper Mgeta)—and the large market town of Dar es Salaam via the national road that links the two cities.

The Uluguru Mountains are the first relief to be hit by humid air masses coming from the Indian Ocean, and from a rainfall perspective, there is a very marked opposition between the two hillsides. While the eastern side receives the first rains (orographic effect), the western side, which is the subject of this study, is downwind and, at comparable altitudes, receives much less rain. However, it receives more than 1000 mm of rain at around 1200 m in altitude, 1300–1400 mm at Tchenzema, at around 1700 m in altitude (Fig. 6.2), and much more on the high hillsides and summits.

The region studied (part of the Upper Mgeta) is thus characterized by significant altitudinal tiering and a marked climatic gradient: the higher one climbs in altitude, the more the rainy season is prolonged and the more precipitation volume increases, while temperatures drop, thus lengthening the plant growth cycle, particularly that of corn (Fig. 6.3).

This relatively generous rainy season is conducive to multiple crop cycles in the same year, especially since the development of irrigation has helped, as we shall see, further increase the benefits offered by the climate.

*Corresponding author: hubert.cochet@agroparistech.fr

©2024 CAB International. *Agrarian Systems and Climate Change: Journeys of adaptation in the Global South* (eds H. Cochet, O. Ducourtieux, and N. Garambois)
DOI: 10.1079/9781800628137.0006

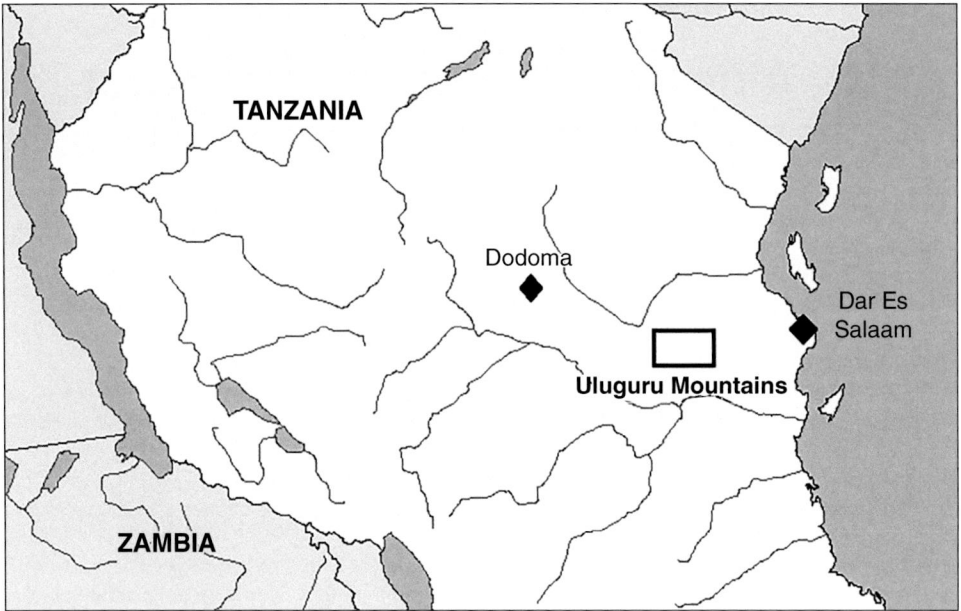

Fig. 6.1. Uluguru Mountains.

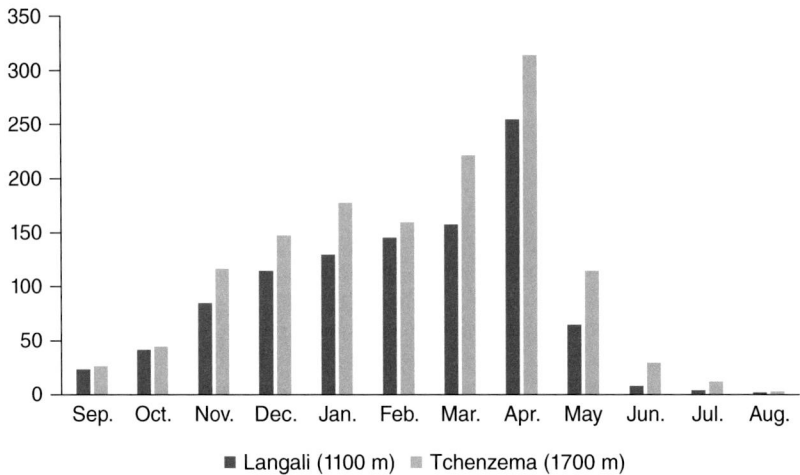

Fig. 6.2. Rainfall distribution and altitudinal climate gradient (1960–1977 averages). (Data: Coniat, 1993; graph: authors.)

6.1.2   Farmers talk little about climate change

The possible changes in climate in this region of the world have already been discussed in previous chapters (Chapters 2 and 3). Forecasts concerning global precipitation volumes are uncertain, but the accentuation of violent episodes is surer,

which, in mountain regions as in the flood-prone areas studied in Chapter 3, could very well aggravate farmers' difficulties. However, the climatic data to which we have had access are quite old (see above), and more recent series, which would have allowed for a comparison with the older ones, could not be consulted, assuming that they truly exist for the region concerned.

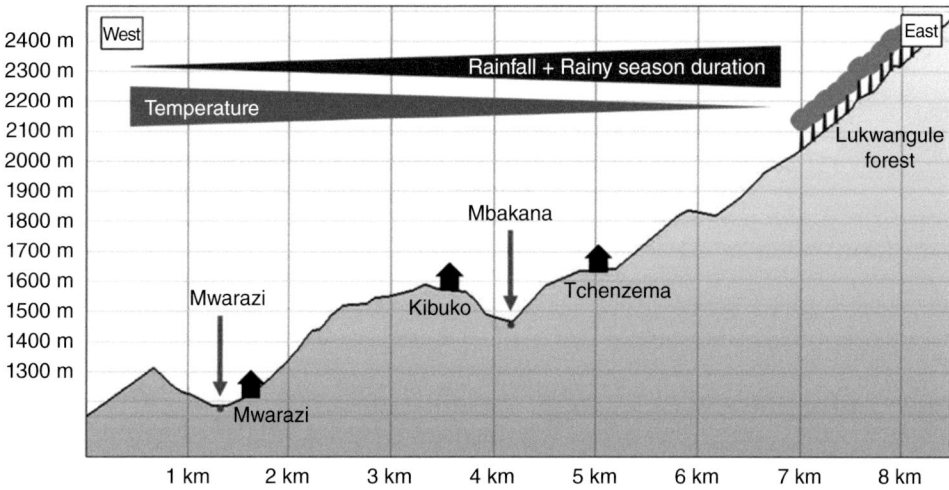

Fig. 6.3. Temperature and rainfall gradient along the leeward (western) side of the Uluguru Mountains. (Graph: Hartog, 2016.)

The function of the water tower in the massif's forest summits, so important for supplying the irrigation network in the dry season, does not seem to be threatened a priori. On the other hand, if the Uluguru Mountains do not escape the increased frequency and intensity of extreme events, they will be particularly affected, due to their slopes, by the risks of erosion by runoff and the ever-present risk of landslides.

The likely increase in temperature will undoubtedly lead to the elevation of ecological stories, moving crops (and pest communities) upward. The lower levels of the leeward (western) side could then suffer from a decrease in precipitation or an increase in evaporation, both of which would cause a greater relative dryness.

But the farmers we met did not spontaneously talk about "climate change"; the few testimonies we gathered on this subject mainly mentioned, as in many regions, the problem of the (increasingly?) erratic arrival of rains in October.

6.2 Cultivating Diversity: The Progressive Intensification of the Luguru Agricultural System

The Uluguru Mountains, which are limited in size and relatively isolated in the lowlands of eastern Tanzania, were settled late in the 18th

century. The stages of this settlement and the associated (matrilineal) family structures, as well as the resulting land tenure structure, have been analyzed in detail by Jean Luc Paul (Paul, 2003).

Initially cultivated with slash-and-burn agriculture during the pioneer phase, the hillsides of the Uluguru Mountains have undergone profound and rapid changes in cultivation practices and environmental management. This region thus provides a spectacular example of the gradual intensification of cultivation practices, linked, of course, to the rapid increase in settlement and the early closing off of the pioneer front, but also integration with regional trade, particularly through the development of market gardening.

6.2.1 From slash-and-burn agriculture to plowing and the progressive planning of hillsides

The range of cultivated plants was initially dominated by finger millet and sorghum, cowpea (*Lablab niger*), and root and tuber plants (taro, yam) and, probably from the 18th and 19th centuries onward, was complemented by the addition of plants of American origin, notably corn, beans (*phaseolus*), and cassava (Paul, 2003). The tools used corresponded to slash-and-burn cultivation: ax (*hwago*), brush hook (*sengo*), sowing stick (*muhaya*), and a type of wooden hoe (*chibode*) used for weeding in the rainy season (*idem*).

It is likely that *Luguru* farmers took advantage of the diversity represented by the different ecological stories to spread their fields over the entire altitudinal range available to them from the beginning of the massif's colonization. The main cropping cycle was generally based on the rainy season, with planting occurring at the beginning, in October–November. At the highest elevation, corn was planted at the end of the rainy season, in May. At this altitude, prolonged rains, fog, and residual soil moisture allowed corn to complete the beginning of its cycle in May and June, at the start of the "dry" season, while lower temperatures extended the cycle's duration until February–March of the following year (Fig. 6.5). The variable length of the vegetative cycle, depending on altitude, temperature, and the varieties used, already allowed farmers to spread out cultivation cycles within the work calendar and thus increase the area cultivated per worker, and thereby overall labor productivity.

As soon as most of the hillsides were cultivated using slash-and-burn techniques (with the exception of massif's summit area, above 1800–2000 m in altitude, which was maintained under forest cover), and as the growing population and new arrivals reduced the space available per laborer, forest recrudescence periods were rapidly shortened, and the secondary forests separating two cultivation phases gave way to shrubby and then herbaceous recrudescence. Important changes in the way the environment was exploited accompanied the vegetation cover's evolution, in terms of weed control on the one hand, and fertility reproduction on the other.

6.2.1.1 Stage 1: from slash-and-burn to clear-and-burn on herbaceous fallow land

The first stage can be characterized by a transition from a slash-and-burn system to clear-and-burn agriculture on sward. From the beginning of the 20th century, the first missionaries' testimonies show a largely deforested landscape dominated by crops and herbaceous fallow land (Paul, 2003). J.-L. Paul relates the evolution of the crop mix used by farmers to the sward's evolution and the increased difficulties that it brought about for weed control. Thus, finger millet, a short, fine-strawed cereal that is very sensitive to weeds, was quickly replaced by sorghum (probably in the 19th century) before the latter gave way to corn (Paul, 2003). In addition, farmers likely abandoned their wooden hoes (*chibode*) for iron ones (in the late 19th century), which were much better adapted to controlling herbaceous sward (*idem*).

Regarding reproducing the fertility of cultivated fields, as the role played by tree regrowth diminished (reduction in the volume of biomass cut and burned), farmers set up a particular system of soil preparation associated with clearing and burning, which the oldest people we interviewed are still able to describe. This work consisted of first cutting the herbaceous vegetation and possibly the shrubby regrowth with a brush hook (*sengo*) during the months of August and September, 2 months before the start of the rainy season. The biomass was left to dry, then burned over the entire field during October, with a few meters of the field edges being cleared beforehand to prevent the fire from spreading to neighboring fields. Once this task was completed, a localized hoeing of the soil allowed the roots of the remaining clumps to be removed, biomass that was then gathered in order to form windrows arranged along the contour lines[2], every 20–30 m. This practice made it possible to both limit erosion, due to the swaths created (and recharged each year), and produce a sort of compost spread out every 5 years or so over the entire field after the swaths were destroyed. The organic matter was thus partly preserved and not totally burned. Sowing was then done without any further tillage. Manual weeding with a hoe was necessary during the month following the first rains' arrival.

In addition, sheep and goat breeding, which developed during this period, allowed for the activation of a fertility transfer from the grasslands to the fields directly adjacent to the house. In fact, dry manure accumulated by night-time resting herds near (or inside) houses was regularly "swept" toward the banana plantation and fields located further down from the house, which thus benefited from this new fertility contribution.

The transition from a slash-and-burn system to a clear-and-burn one (Fig. 6.4) thus marks the first labor intensification phase for this agrarian system.

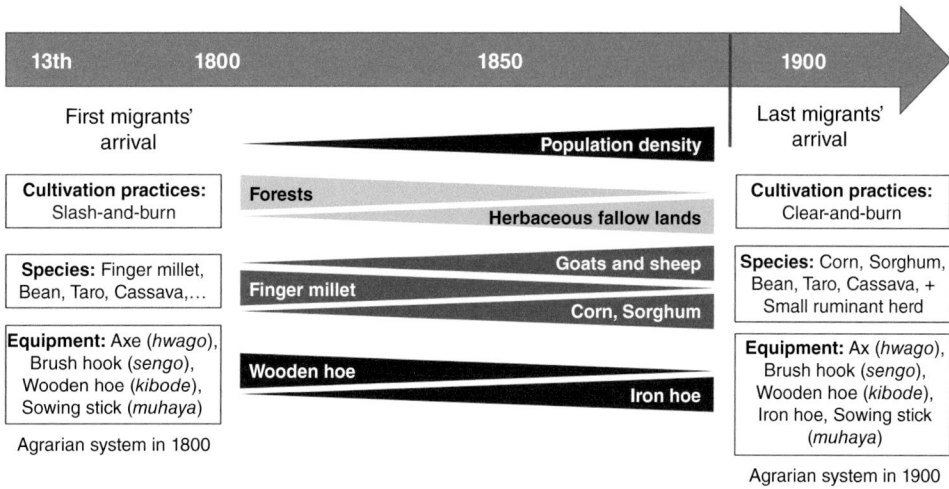

Fig. 6.4. Summary diagram of the main transformations over the 19th century (Diagram: authors (adapted from Paul, 2003).)

6.2.1.2 Stage 2: field and tillage planning

A second stage can then be highlighted: the generalized practice of tillage in the form of real plowing, associated with the establishment of continuous cultivation and the progressive spread of different hillside planning practices with *ladder terraces* (*matuta ya ngazi,* see below).

While crops of American origin—corn, *Phaseolus* bean, sweet potato, and cassava—occupied more and more space in the cultivated plant mix, the debut of market gardening at the colonists' initiative (but initially forbidden to the natives [Coniat, 1993]) as well as the development of irrigation[3] makes it possible to cultivate all ecological stories in the dry season (not only the highest, which already benefited from more abundant rainfall and atmospheric humidity) and thus multiply crop cycles on the same field.

Although the colonial context[4] was not conducive to these transformations and contributed to the deterioration of natives' living conditions, the subsequent relaxation of colonial pressure[5] and then independence allowed these new elements to contribute decisively to the agricultural system's evolution. It was during the 1950s and 1960s that market gardening began to take on real importance, while the market developed, and irrigation networks continued to expand. As the first roads opened up the region, the marketing of vegetables was taken over by cooperatives (Uluguru Farmers Cooperative Association, or UFCA) in the 1950s, then by private traders. The main vegetables grown at that time were cabbage, cauliflower (both in the colder ecological stories), and more rarely, onions and potatoes. Generally, cabbage and cauliflower are transplanted at the end of the rainy season, after the harvest of food crops such as corn and beans, and irrigated in the second part of the cycle. At high altitudes, cabbage and cauliflower can be combined with corn during the first phase of corn development (Fig. 6.5).

The 1950s also saw the development of temperate climate fruit trees (plum, peach, apple, pear), which had been imported into the region by settlers at the beginning of the century but whose planting by natives was strictly forbidden throughout the first half of the 20th century. As soon as the colonists left, farmers took advantage of their new freedom to plant these trees around and within their fields.

At the same time, while population continues to increase during this period, the reduction in farms leads to a decrease in multi-year fallow lands and, simultaneously, small ruminant herds.

As for soil preparation, one moves from shallow, flat tillage after clear-and-burn, with partial composting of organic matter (*above*), to ridge tillage, similar to true plowing with total burial of organic matter. This evolution requires significant labor intensification, which is justified by populations' need to compensate for a

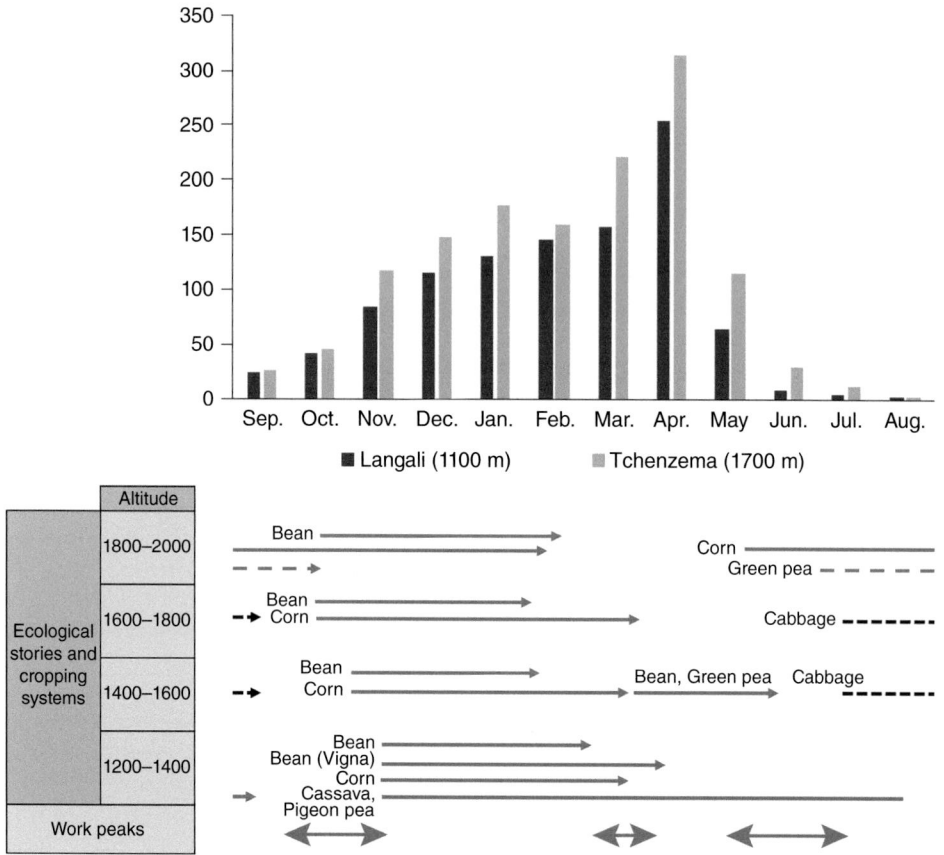

Fig. 6.5. Rainy season and work calendar in Mount Uluguru (1950–1970). (Chart: authors.)

decrease in soil fertility, following the increase in the frequency of cropping cycles.

The Uluguru Mountains, identified in colonial times as an already "overpopulated" region and plagued by what was declared to be intense erosion, were the target of a particularly authoritarian erosion control policy that aimed at planning the hillsides into radical terraces. We will see in paragraph 6.4 that this policy, far from achieving its stated objectives, ended in abject failure and was quickly abandoned. However, the hillsides of the Uluguru Mountains were still carefully terraced and staircased in different ways, but by the farmers themselves and with a wide variety of planning practices (*see* Photos 6.1 and 6.5). It was at this time that farmers began to use livestock waste in fields farther away from dwellings, especially on market gardens, and pig farming developed.

6.2.2 Expansion of irrigation, spread of market gardening, use of chemical inputs, and gradual filling of the crop calendar

6.2.2.1 Development of irrigation networks

The irrigation network has continued to develop from the 1950s to the present day, especially since the 1990s. The development is such that today it is difficult to find a farm without access to irrigation.

The development of irrigation has manifested in different ways:

- First, there was a significant increase in the number of canals from main rivers and their tributaries, some of which can be several kilometers long. In the early 1980s, agricultural surveys conducted on about

100 production units located in the Nyandira village territory indicated that the network already covered 47% of fields and 87% of farms (Paul, 1985, p. 10). As an example, Coniat already counted 15 water intakes, or 30 connections (one canal departure on each bank for each intake) on the Mindu torrent at altitudes ranging from 1900–1450 m (Coniat, 1993, p. 163). The secondary canals consisted mostly of simple streams running down the slope perpendicular to the primary canal to serve the fields on either side. The watering of each plant (cabbage, for example) was then done with the help of a plate by drawing from micro basins set up here and there at the edge of the field.

- More recently, the development of hosepipe irrigation, to be used instead of gravity-fed irrigation combined with the use of plates (*see* Photo 6.2). The hosepipe allows for a more precise and economical use of the resource, allowing a greater number of farmers to irrigate at the same time, from the same primary canal. Connecting the hoses directly into the primary canal also avoids damage from runoff into the secondary canal. The purchase of an irrigation pipe, however, depends on farmers' investment capacity and many continue to irrigate "at the plate."
- Finally, the organization of irrigators' groups and a tendency to monetize access to water. Indeed, it is common for those who took the initiative to dig a new canal to form a group and demand annual or permanent membership fees from those who did not participate in the construction of said canal but wish to use it. This monetization of water rights concerns mainly the most recent canals, but some people still claim rights for older canals built by their parents.

6.2.2.2 Spread of market gardening

As population density continued to increase and farm size declined sharply with the settlement of the new generation, Uluguru Mountain farmers turned decisively to vegetable crops, whose higher income per unit area was able to sustain and reproduce their farms.

In the 1960s and 1970s, the range of vegetable crops expanded with the arrival of new vegetables (lettuce, spinach, Chinese cabbage, cucumbers, peppers), which were integrated into the Luguru agricultural system (though still in a limited capacity), showing the farmers' desire to diversify their source of income. While the first vegetables (cabbage and cauliflower) were integrated into the food crop systems that were previously predominant (*supra*), the new vegetables were grown in the houses near the vicinity on dedicated irrigated fields and more and more often planned in true terraces. In the 1980s, J.-L. Paul clearly distinguished this new "market gardening system" from the "mixed systems" that had appeared earlier (Paul, 1985, 1988). At the time, agricultural surveys conducted in Nyandira (*supra*) indicated that only 7% of the fields in the sample studied were cultivated in this way, concerning 25% of the farms (Paul, 1985, p. 12), whereas cabbage and cauliflower were present in 94% of the production units (*idem*).

Since then, diversified market gardening has become widespread, with new market garden crops that have higher value added and less weight than cabbage and cauliflower (relative to their price), proving to be very attractive despite their sometimes-high production costs. The cultivation of tomatoes has developed during the 2000s and has been very successful. Cabbage and cauliflower still have a prominent place, but they are now grown alongside lettuce, Chinese cabbage, leeks, etc.

6.2.2.3 Fertility reproduction: development of pig farming and use of chemical inputs

However, significant market garden crop exports outside the agricultural system weigh on soil fertility, in a context where vertical and horizontal fertility transfers within the system have considerably diminished (decrease in fallow land and wooded areas, regression of small ruminant farming). Also, farmers are rapidly integrating different chemical inputs on fields dedicated to market gardening, while increasing the areas devoted to these crops within their cropping pattern.

Chemical fertilizers became available in the region in the late 1960s. Synthetic fertilizers were made available free of charge by the government during the Ujamaa era, but they quickly became expensive in the following decades. They were mainly used for market garden crops, although some farmers were already beginning to use them for food crops as well.

The development of market gardening, made possible by chemical inputs, is also accompanied by an increase in the number of pigs (*see* Photo 6.3). The presence of pigs, which until then seemed to have been discreet, began to develop in the 1960s and 1970s, notably through the use of corn "bran" imported from flour mills in the surrounding regions (Morogoro), combined with various products from the farm (banana and cassava leaves, taro, spoiled fruit, etc.). It is directly associated with the cultivation of market gardening, which has significant manure requirements. The development of livestock farming allows for an increase in the input of organic matter, with bran imports from the plain constituting a transfer of biomass from outside the region. Thus, 62% of the farms in Nyandira village were already raising pigs (1 or 2, 1.3 on average) at the beginning of the 1980s, with pigs playing a major role in fertilizing market garden crops (Paul, 1988, p. 31 and p. 42). At the same time, small ruminant livestock production had been declining with the gradual disappearance of intercropped fallow land and rangeland, a decline that has been partially offset more recently by the development of dairy goat production resulting from dissemination programs implemented in the 1990s.

Thus, the use of organic manure and synthetic fertilizers has increased considerably over the past 25 years and is no longer limited to vegetable crops, but also extends to food crops on each farm. While in the 1980s, J.-L. Paul estimated that 28% of farms used fertilizers on corn crops, "occasionally and on small areas" (Paul, 1988, p. 16), we now estimate that 90% of a sample of 64 corn fields analyzed, all crop associations included, were fertilized. These fertilization practices for food crops concern all types of farms, from the smallest to the largest. This increase in the use of inputs is primarily a reflection of soil fertility degradation, which is mainly due to market gardening demands, crop exports, and the lack of organic nutrient recycling.

During the 1970s, the first plant pest control products also appeared on the market in the region. These products seem to have been used first in the higher areas, where the density of vegetable crops and land use was greater. The lower areas began to use these chemicals later, in the 1990s. Their use is now systematic in market gardening, and even excessive on tomatoes, to the point of serious concern to development organizations working in the region.

In addition, the generalized use of irrigation, development of market gardening, and improved market access have made it possible to multiply crop cycles and adjust them over time, including food crops, thus paving the way for better use of the family labor force (Fig. 6.6), in a context of dwindling peasant tenure. This diversification has also allowed for better adaptation to annual variations in market prices[6].

6.3 Accessing Different Ecological Stories, Maintaining and Increasing Crop Diversity, Accessing Inputs and Markets: The Keys to Success

All farmers in the region grow food and market garden crops (with a ratio that depends on their cash flow and access to irrigation) on fields scattered throughout the different ecological stories and at the same time invest in livestock according to their income. The types of farms established in this analysis are therefore distinguished mainly by their access to land, irrigation, and the market.

6.3.1 Increased differentiation

Although the farms are on average very small—a few acres—there are important differences between the smallest, less than 1 acre in size and with little access to irrigation, and the largest, which can be as sizable as 10 acres, most of which are irrigated. In addition to the inequalities in access resulting from the order and conditions in which the different lineages arrived in the region during the pioneer phase, there are those inherent in the processes of differentiation that have become more pronounced since the second half of the 20th century and that are linked above all to conditions of access to irrigation, market gardening, and the market (proximity of roads in particular). Although the canal

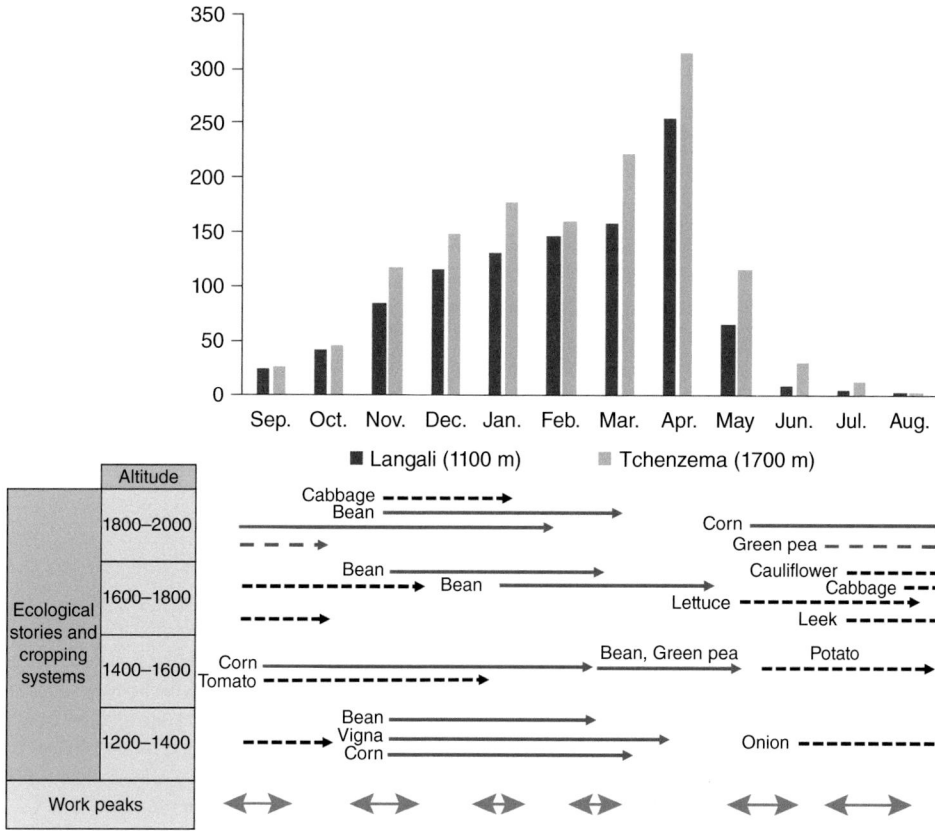

Fig. 6.6. Rainy season and work calendar in the Uluguru Mountains today. (Chart: authors.)

network has developed considerably over the last 20 years, facilitating water access for many families, some areas still lack access to irrigation or are located at the end of the network and deprived of sufficient water. In addition, the pressure linked to this resource is now leading to an increasingly frequent monetization of water access. Thus, a young couple that has received fields of sufficient quantity and quality, for example, 2–3 acres of which at least one is irrigated and located on the edge of the road, can enter into a process of capital accumulation and gradually increase their surface area by purchasing new fields. For other families, who have no access to irrigation and too little land, the situation is quite different.

In addition, with the rise of cash crops, but also the increase in land pressure and difficulties in accessing land, the gap has widened between families with large farms oriented toward cash

crops, unable to carry out the amount of work required on their own, and families with insufficient land to meet their needs, forced to sell part of their labor force to ensure their survival. This movement was described in 1985 by J.-L. Paul, who explained that the system of buying/selling the labor force mainly concerned the largest and smallest farms, while medium-sized farms remained faithful to the system of mutual aid that had prevailed until then. In a sample of about 60 farms, 35% bought labor, 11% sold it, and 87% helped each other (Paul, 1988). The analysis of the current situation tends to show that this commodification of the labor force has increased today so that the majority of farms now employ labor at least once a year. The purchase of labor does not only concern large farms but also concern medium-sized farms and even small farms. It is not uncommon for farmers to buy and sell labor in the same year. Finally, many workers, especially young people, live mainly from the daily sale of their labor for field work or as porters.

6.3.2 Simplified production unit typology

6.3.2.1 Very small farms with little access to irrigation and the market, forced to sell part of their labor force

This first category includes the smallest farms (0.5–1.5 acres), which have very limited access to land, but also often to irrigation and the market. These farms are in a very fragile financial situation, which makes vegetable production (demanding in inputs) difficult to implement. These farmers are sometimes forced to work as day laborers for other farms, not only to meet the household's needs for a significant part of the year but to purchase inputs (seeds, fertilizers, plant pest control products) and pay the canal access fee. This results in a delay in farm work; cabbage will thus be sold at its lowest price.

The cropping systems implemented depend, of course, on the altitudinal story. The farmer always carries out a food crop cycle during the rainy season, which most often consists of a corn crop in association with beans, for example, at the highest story of the profile (1900 m) (Fig. 6.6). During the dry season, the farmer can produce, for example, legumes (beans, peas) or vegetables (cabbage, cauliflower, lettuce, Chinese cabbage, carrots, etc.) on a small area of 0.3 and 0.2 acres located a little further down (1700 m): the farmer sows corn and beans at the beginning of the wet season (October/November) and then, during the dry season, irrigates beans and cabbage, with fertilizers and insecticide treatments.

These farmers generally have no or very few animals. Some of them may be able to fatten a pig, but they are often forced to sell it early, at a disadvantageous price, due to lack of cash. The pig is fed mainly on grass and corn bran from the farm. Additional bran may be purchased depending on cash flow, at a rate of two buckets per month. The recovered manure is used in the fields, especially for cabbage.

Equipment is reduced to two hoes, an axe, and a machete. Lacking sufficient investment capacity, these farmers are forced to rent a portable sprayer and an irrigation hose by the day.

Livestock buildings are reduced to covered pens rebuilt each year with wood harvested from the forest.

The annual income of this type of farm is very low (200,000–500,000 Tanzanian shillings [TZS], or only 40–100 euros/laborer/year) and they could not survive without the work the farmer and their family are forced to carry out on the outside. The reasons for this are the farm's small size, access to only one or two ecological stories, and the limited range of possibilities available to the farmer in terms of cropping pattern. As a result, only about 100 days of work can be devoted to the farm each year, which illustrates the "forced" underemployment of the family labor force.

6.3.2.2 Small farms with good market and irrigation access

This second category includes farms with 2 or 3 irrigable acres that are oriented toward vegetable crops. Typically, these farms produce at least two or three different vegetable crops, which allows them to achieve a much higher income than the previous category. However, these farmers take significant risks, as vegetable crops are input-intensive and market prices are volatile. In fact, these farms often operate "like an accordion," with high economic results in good years and potentially catastrophic results in bad years, forcing farmers to decapitalize.

On 2 or 3 irrigated acres, spread over three different ecological stories, it is possible to carry out two crop cycles on all the fields, giving priority to high-value crops such as tomatoes, onions, and cabbage. For example, at around 1900 m in altitude, corn and beans are grown together in the rainy season, followed by cabbage in the dry season; at 1600 m in altitude, corn and tomatoes are grown together, sown in September, and irrigated while waiting for the rains, followed by beans or peas at the end of the wet season; and finally, at around 1400 m in altitude, corn is associated with beans in the rainy season, and onions are irrigated afterward (Fig. 6.6).

Fertilizers and plant pest control products are widely used, particularly for market gardening, but also for food crops. This type of production unit has all the equipment necessary for its operation: two hoes, an axe, a machete, a 25-m irrigation pipe, and a portable sprayer. The livestock buildings are reduced to covered enclosures built with wood harvested from the forest.

The two fattening pigs are purchased at 3 months of age and sold at 10–12 months of age. During this period, they are fed mainly with corn bran, bought every week at the market, as well as with farm by-products.

In addition to the family labor force (usually a couple), a few days of outside labor are purchased at peak work times, which can be crucial for increasing the area cultivated at the appropriate time, beyond what the farmer and his wife would be able to do.

Farm income is then clearly higher, around 500–600 euros/laborer/year, and in good years sufficient for the farmer to avoid outside work. Not only does access to different ecological stories allow for better use of the family's labor force (more than 200 workdays per year are devoted to this), but crops with higher value added (tomatoes, in particular) allow for greater value added and remuneration of labor. However, in the event of a poor harvest or a drop in prices, the farmer is often obliged to decapitalize, or, if that is not enough, to work outside as a day laborer. These farms remain fragile but have the potential to achieve good economic results, which allow them, in some cases, to invest gradually in the land and expand.

6.3.2.3 Medium-sized, intensive farms, oriented toward market gardening

This category of farms includes larger units (about 5 acres), most of which are irrigable (80%) and have access to a wide range of climatic conditions spread out along the hillsides. This dispersion of fields along the altitudinal gradient, far from being a handicap, allows them to produce diversified vegetable crops in quantity—generally between four to six different vegetables—which enables them to reduce the risks involved. These farms also have larger livestock operations and generally have no difficulty in obtaining all the equipment they need to operate, including hoes, axes, and machetes, as well as two 50-m hoses and two portable sprayers. The livestock buildings are permanent, built from more durable materials.

The two-family workers are supported by day laborers throughout the year, for the equivalent of about 50 days of work, especially during plowing and sowing, a key period in the work calendar.

While a rainfed corn cycle is always carried out on fields located at altitude, the farmer directs the majority of his fields toward irrigated market gardening, for example: dry-season cabbage behind rainfed corn + bean around 1700 m; dry-season lettuce behind rainfed bean around 1600 m; dry-season potato behind corn and tomato (irrigated at the beginning of the cycle) around 1500 m; finally, lettuce and bell pepper under irrigation in the warmer story around 1400 m in altitude. These farms systematically use synthetic inputs.

These farmers have been able to invest in both pigs (breeding and fattening) and dairy goats, which are reared tethered. During lactation, they receive a daily feed supplement in the form of corn bran. Since the corn bran is produced on the farm in insufficient quantities, the farmers are obligated to buy it at market.

The access to a wide range of crops and the less restrictive surface area of these farms allow them to make much better use of the available family labor force (about 400 days of family labor), the cultivated area being expanded by hiring day laborers at peak work times (about 50 additional work days). The annual agricultural income is then established at more than 1500 euros per family worker.

6.3.2.4 Large multipurpose farms

Some farms are much larger, about 8–9 acres, spread over a wide range of ecological stories, and three-quarters of them are irrigable. These farms are able to diversify both their vegetable and non-vegetable crops, which gives them an advantage in terms of income and risk mitigation. On the most accessible and irrigated fields, vegetable crops are preferred, while the more remote fields are reserved for pulses and food crops.

Several livestock are also present: some pigs for fattening and a sow for breeding, as well as a small herd of dairy goats. In addition, the poultry farm (always intended for self-consumption) is of a larger size thanks to the large areas of corn that allow for a surplus of grain used to feed the chickens. However, the corn bran produced remains insufficient and needs to be supplemented by purchases at the market. The farm thus benefits from a larger quantity of animal manure.

All of these activities keep the family labor force busy throughout the year (about 500 workdays for a farming couple) and thus increase overall labor productivity (value added per worker on the farm). In addition, it is necessary to employ day laborers during peak periods, for a total of about 200 workdays, which is a crucial complement to increase the cultivated area.

The equipment on these farms, although still based on manual tools, is more complete, for both crops (portable sprayers, hoses) and livestock (permanent buildings, including for poultry).

These farms are thus characterized by a wide variety of income sources from both crops and livestock. The income per family worker reaches 2000 euros/year or more.

6.3.2.5   Medium-sized farms, located at lower altitudes and with little irrigation access

A number of farms, located at lower elevations, have less access to irrigation and are mainly oriented toward cereals, legumes, tubers, and perennial crops (fruit trees and sugarcane). On the 4–5 acres available to them, these farms generally diversify their non-vegetable crops, taking advantage of the different ecological stories and the different crop associations possible. In the warmer stories, corn, beans (*Phaseolus* and *Vigna*), pigeon peas (*Cajanus cajan*), and cassava are often grown in combination. In the higher altitudes, the farmer favors corn associated with beans and sometimes squash. Peas are also often grown, especially during the dry season. A small portion of the land (less than 20%) can be irrigated—these farms are often at the end of the network—and is reserved for market gardening, particularly tomatoes (associated with corn).

Except for irrigated tomatoes, very few chemical inputs are used on this type of farm, as pig manure makes it possible to do without synthetic fertilizers.

Labor productivity and income are lower than in the three types previously described (income is about 450–500 euros per worker), due to less access to higher ecological stories, water, and higher value-added crops (excluding tomatoes). These farmers take relatively little price risk, since their production costs are lower, and the prices of non-vegetable crops are generally more stable. If the predicted temperature increases

occur with force in these already warmer parts of the region, and in a context where irrigation water is likely to be increasingly in demand from neighbors in the upper stories, it is probable that these farmers will be more vulnerable in the future.

6.4   Planned Peasant Farming, Erosion Control, and Adaptation: Lessons from History

As we have seen, the Uluguru Mountains region was identified from the colonial era as an "overpopulated" region and prey to intense erosion. It was therefore the target of a particularly authoritarian erosion control policy, the main tool of which was to plan the hillsides into radical terraces. Insofar as the current policies advocated in terms of adaptation to climate change (Chapter 8) focus on the fight against land degradation (erosion, overgrazing, deforestation, salinization, reduction in organic matter, soil impoverishment, etc.), the example of the Uluguru Mountains is rich in lessons. It illustrates, as do many other situations in Africa and elsewhere, the inadequacy of government programs and projects for local contexts and peasant practices and logic. Far from achieving the stated objectives, the remedies have too often been worse than the disease.

6.4.1   The example of the ULUS program to combat land degradation, implemented by colonial authorities

Through its authoritarian implementation by the colonial power, the reactions that it aroused on the locals' part (including violent ones), and its hasty abandonment, this development project left deeper traces in the collective memory than in the landscape it was supposed to remodel durably. P.H. Temple carried out a critical analysis of this project, published as early as 1972 in the physical geography series *Geografiska Annaler* (Temple, 1972), on which this section is based.

The erosion control policy implemented through British colonization of the Uluguru Mountains in the early 1950s was one of the most ambitious ever carried out at that time.

Prior to this, the forest reserve on the summit plateau of the Uluguru Mountains had been set up by the German administration, its boundary marked, and the fields and houses within it moved. While today there is virtually no doubt in the interest of this protection for safeguarding the water tower created by the massif's summits, the consecutive increase in anthropic pressure in the surroundings led to a decrease in the time of fallow land on the steep slopes of the hillsides surrounding the protected massif, as Savile pointed out very early on (1945–1946, cited by Temple, 1972). It then contributed to precipitating the crisis of the slash-and-burn system.

The development of perennial crops (coffee) and digging of contour canals parallel to the contour lines was then advocated to control runoff, as in neighboring Burundi[7]. The ineffectiveness of these measures led to the proposal to empty the region of its population in order to turn it into a wilderness reserve (Temple, 1972, p. 115). Did the colonial sisal plantations in the lowlands not have a high need for labor at the same time?

In the late 1930s, the first "bench terraces" were established in 1936–1937 (Temple, 1972, p. 114), paving the way for the development of the ULUS (Uluguru Land Usage Scheme), which was implemented between 1945 and 1955. While the ban on burning was reinforced, and the steepest slopes were put under protection, the funds, released in 1947, were intended primarily for the construction of "bench terraces," which were supposed to be the only effective protection against runoff and landslides (see Photo 6.4). From 1953 onward, each household was required to build 500 m[2] of terraces per year, while coercion measures were strengthened (Temple, 1972, p. 117). The plan also forbade certain drainage practices, such as drains or stony ditches, which would have allowed excess water to be evacuated while limiting land loss as much as possible.

In the beginning, the Uluguru peasants did not mind, being more concerned about the ban on burning, the fencing of the steep slopes, and the rumors of forced migration to the lowlands. However, they rebelled in 1955, and the scale of the uprising led the British authorities to wipe the slate clean[8].

The program's resounding failure to combat erosion and land deterioration was analyzed by several authors in the years following its abandonment. The techniques proposed were inadequate, particularly that of "bench terraces," which were very difficult to build; had such fragile and erosion-prone banks; and dug up the soil's deeper and less fertile horizons while considerably increasing the risk of landslides (see Photo 6.4). Erosion had long been presented as alarming, although no evidence was provided (Temple, 1972, p. 115) and no diagnosis of farming practices was undertaken, particularly with regard to the ridge plowing methods (*supra*) or the "ladder terraces" that farmers had already begun to build in the 1930s (Coniat, 1993, *see* Photo 6.1 and 6.5)[9].

The population density was also grossly overestimated: 290 inhabitants/km[2] in 1945—the 1967 census gave the figure of 119 inhabitants/km[2]—to support the theory of imminent catastrophe (Temple, 1972, p. 116). The customary authorities were not consulted.

6.4.2   Erosion control and hillside planning: the Uluguru Mountain peasants' way

The colonial authorities' obvious failure in terms of erosion control did not prevent the hillsides of the Uluguru Mountains from being meticulously sculpted with hoes and planned in terraces and staircases of various types by the farmers themselves and in a great diversity of practices, depending in particular on the hillside's slope.

6.4.2.1   Ridge plowing and hillside planning

Thus, all fields are now plowed and cultivated in ridges parallel to the contour lines (*see* Photo 6.5). In all cases, the fields, which are left fallow for a few weeks or months between two cropping cycles, are initially covered with grass and residues from previous crops, with the ridges from the previous cycle still quite marked. Therefore, two main types of plowing can be distinguished, depending on the hillside's slope.

The first type corresponds to hillsides with a relatively moderate slope. Starting at the bottom of the field and working toward the top, farmers use a hoe to remove weeds that have grown on the old ridge and rake them down to the area where the previous year's ridge was separated from its neighbor. In the case where cumbersome crop residues are still present (corn stovers, for example), these are cut or torn off beforehand, then laid down in the old furrow. The old ridge is then broken up and the turned-over soil thus covers the biomass in the old furrow, which then becomes the year's future ridge, while this hoeing leads to the formation of the new furrow on the site of the old ridge. The ridges thus shift from year to year toward the bottom of the field. This type of tillage has often been accused of being responsible for land progressively moving down in this region, as in other neighboring mountainous regions (Cochet, 2001)[10]. This is not true, as this process results in the formation of a bulge further down from the field, marking the separation from the neighboring field below and thus contributing to the formation of pseudo-terraces shaping the hillside (*see* Photo 6.6). The stabilization of the steep (sometimes vertical) bank that separates the two fields is essential. Sometimes vegetated—a practice that is still relatively uncommon in the Uluguru Mountains—this bank is reinforced here by the fact that the bottom bulge of the field is also cultivated in ridges, but this time perpendicular to the previous ridges, that is, this time in the direction of the slope, which is very shallow or even non-existent on this ridge (*see* Photo 6.6). This change in tillage orientation has a double advantage: it avoids creating a ridge on the edge of the bank, which would inevitably be unstable; the burial of organic matter allows the end of the ridge to be "reinforced" and limits the descent of soil to the neighbor below.

The second type is reserved for the steepest slopes and shapes them into real stairs. In this case, the grass (and any crop residues) is first scraped with a hoe and brought back to the lower step, forming a narrow swath. Then the edge of the upper step is cut away, as well as the almost vertical bank that supports it, and the soil detached, thus being used to cover the previously formed swath of biomass and form a new ridge. In this configuration, part of the old step's vegetation is not dislodged and remains rooted. This results in the "cut" steps' stronger resistance to erosive agents, as part of the soil remains in place. These characteristics make ridging possible on the steepest slopes, limiting erosion phenomena. Possible perennial crops are also spared by this tillage (banana and taro, for example, as can be seen in Photo 6.7)[11].

In the stoniest fields, stones are collected and then left at the sides of the fields (and sometimes in the center), so as to form drains for excess water drainage, limiting soil losses; they can also be used to reinforce secondary channels (*see* Photo 6.8).

In both cases described above, one of plowing's functions—an essential one—is to bury organic matter, both that constituted by crop residues and that which has developed spontaneously during the crop cycle (weeds) or after harvest. This is the price to pay for maintaining at least a partial organic matter level in the soil, and the addition of animal manure can supplement the process.

6.4.2.2 "Real" or "radical" terraces

Despite the ULUS project's resounding failure in the 1950s (see above), a few subsequent attempts to popularize radical terraces have left their mark, particularly for vegetable crops grown close to houses. The terrace is then built *ex nihilo* and results in the constitution of larger horizontal terraces, but whose width is limited by the slopes' steepness (*see* Photo 6.9). The very steep banks thus constructed remain very fragile and do not fail to be rapidly notched by rill erosion. This is why, contrary to the attempts made during the ULUS project and its aftermath, this type of terrace is observed today on low-slope fields, often near houses or roads, and devoted exclusively to irrigated market gardening rotations.

In addition to the aforementioned disadvantages of this type of terrace, it should be added that these permanent structures make it difficult to bury organic matter, unlike the ridge plowing mentioned above. At the end of the dry season, when it is time to plant off-season irrigated crops, farmers are forced to burn the organic matter that is present, as the grass has been removed beforehand. This is the only way to get rid of it (*see* Photo 6.10). Secondly, farmers carry out surface tilling to bury the ashes. In this case, phosphorus and potassium are returned to the soil, but carbon and nitrogen are burned. Synthetic fertilizers are then used. If plowing takes place at the end of the rainy season, and the mobilized biomass is not too voluminous, farmers gather it at the edge of the terrace, then cover it with soil. In addition to burying at least some of the biomass, this work restores the terrace edge that may have been eroded by the previous cycle's rains, giving back its horizontal profile thanks to the "framework" formed by the bio-mass piled up along the edge, and thus avoiding a reduction in the terrace's surface area. Finally, in some cases (especially for the widest terraces), farmers can carry out ridging work on each ter-race, perpendicular to the terrace's longitudinal axis, thus regaining the advantages described for ridge plowing.

6.5 Conclusion

Thus, in the Uluguru Mountains, as in many re-gions of the African continent and elsewhere that are characterized by purely manual agricul-ture (no animal or fossil energy) and very low hourly labor productivity, it is the extension of working periods and filling of the work calendar throughout the year that allows real gains in labor productivity (measured per worker and per year). By allowing each family to live on a small-er and smaller land base, this process has also made it possible to cope with sustained popula-tion growth. The diversity of crops grown and their spread throughout the year is, therefore, crucial to produce as much or more on a smaller area and make any increase in productivity pos-sible, even though any form of productive spe-cialization (a single production) would lead to the opposite result: rapid saturation of the fam-ily labor force during the work peak, and chronic underemployment during the rest of the year. Moreover, such development in cash market gardening crops would considerably increase the risks incurred, particularly the price risk

(inputs/production), if it were based on a single production or too few.

This is why access to diversified ecosystems (which requires a certain dispersion of fields rather than their grouping into larger blocks), access to irrigation, and the use of the most diversified cropping systems possible are so important in this context. Provided that access to the minimum means of production is guaran-teed and secure and that access to the market is possible under acceptable price conditions, such diversity of activities, in addition to increases in productivity, is the basis for a reduction in the risks incurred (climatic and agronomic risks, price risks, etc.). It thus increases the small-scale farms' resilience and their capacity to adapt to global changes, including climate change. How-ever, the significant increase in the frequency of phytosanitary treatments raises questions: in add-ition to the increased costs generated by their use (and the inherent price risk), such a dependence on inputs, if it increases further, will not fail to threaten the agricultural system's resilience and undermine the progress made in recent decades.

As for the issues related to fertility reproduc-tion and the "fight against land degradation," the agriculture of the Uluguru Mountains offers a telling example of the efforts undertaken by farm-ers to face up to it and preserve the future, despite the paltry means at their disposal and the past at-tempts at hasty management orchestrated by the public authorities and development projects[12]. Much remains to be done in this area, in particu-lar, to further strengthen the role of livestock (especially pigs) in fertility and tree reproduction (trees still being too few in number), and the stabilization and reinforcement of terraces and parcel boundaries (hedgerows).

Notes

[1] The 2016 fieldwork carried out in the region was greatly facilitated by the contacts previously established in the area and at the University of Morogoro by Jean Luc Paul. We thank him warmly.

[2] This is the "Mgeta system": ridges of grass and weeds placed along the contour lines as a method of conservation is mentioned in the literature on the region's history (e.g. Temple, 1972, p. 113).

[3] Preexisting irrigation. Explorers already mention it when passing through at the end of the 19th century (Coniat, 1993; Paul, 2003).

[4] There was a considerable increase in tax pressure (hut tax and capitation tax), colonial drudgery, land evictions, and the establishment of large colonial farms in this cooler climate region, which were highly prized by settlers (Coniat, 1993).

[5] As well as the departure of the German settlers who had remained in the area despite the establishment of British rule after World War I.

[6] On the other hand, fruit trees, which were developed relatively successfully in the region from the 1950s onward, occupy a limited place in the agricultural system, a limitation that may be related to the land tenure system still influenced by matrilineality (Coniat and Raison, 1997). Few orchards are visible in the landscape, unlike in other neighboring mountainous regions (northern Tanzania, Burundi, etc.).

[7] In Burundi and Rwanda, demographic growth and intensification of practices observed by colonists (reduction of pastures and fallow lands, increase in cultivated areas, multiplication of cultivation cycles, cultivation of steep slopes, etc.) were enough to make erosion's worsening obvious—in their eyes, indisputable. Something had to be done. The "natives" were made to make isohypse "anti-erosion" ditches, ditches by the hundreds of thousands of kilometers (140,000 km of ditches listed in 1945), especially on the pastures— precisely the spaces least subject to erosion—because the work of measuring, staking, and monitoring was easier there and the population's resistance was less strong than on patterned land (Cochet, 2001).

[8] A more precise description of the sequence of events is given by Coniat (1993, pp. 126–128).

[9] "The association of food crops, cassava-sweet potato or corn-legumes (beans, cowpeas, or peas), or sorghum-corn-beans or peas-sweet potato, as well as the beginning of the association with cash crops (corn-cabbage-beans) stabilized the field against sheet erosion thanks to the presence of a continuous plant cover" (Coniat, 1993, p. 131).

[10] In Burundi, too, plowing always begins at the bottom of the plot and progresses upward, throwing the soil downward. As this practice causes the soil to progressively move down the hillside, sometimes referred to as "dry" erosion, farmers have sometimes been advised to plow in the direction of the contour line. However, the latter technique is virtually impossible to implement on steep slopes. To carry out such plowing (or even from top to bottom as some advocate) the position to be adopted would be so uncomfortable that it would be impossible to obtain a complete overturning of the clods and a good burial of the organic matter. Moreover, the witch-hunt made of the bottom-top type of plowing is very unfair. In the case of Burundian farmers, it is the source of the sloping pseudo-terraces that can be observed in many places in the Burundian countryside. It results in the formation of a bulge at the bottom of the field and a relative depression at the top of the field, the result of the last plowing ridge. In a few years or decades, the field's profile is softened, the hillside then being made up of a series of inclined pseudo-terraces (real "progressive terraces") separated by very steep banks. After several generations of plowing from "bottom to top," the banks reach several meters in height and support true horizontal terraces, as can be observed in certain regions. As soon as the terrace becomes horizontal, the direction of plowing is changed every year so that a counter-slope (opposite to the hillside) does not appear, and the field remains flat. This type of plowing, although denounced as the main factor causing erosion, then becomes a true soil conservation and hillside planning technique (Cochet, 2001).

[11] This type of step terracing (matuta ya ngazi) can be recognized in Grant's 1956 description (cited in Coniat, 1993, p. 114). Very few cases of failure were reported with this type of planning.

[12] One element that has favored the hillsides' gradual planning is the low impact of villagization in this region of the Uluguru Mountains, which would have undoubtedly caused serious disturbances if it had been imposed here as it was in other regions of Tanzania (see Chapter 4).

References

Cochet, H. (2001) *Crises and Agricultural Revolutions in Burundi*. INA P-G/Karthala, Paris.

Coniat, D. (1993) *Les Transformations de L'espace Rural, Des Activités Agricoles et de L'environnement Sur Le Terroir Du Haut-Mgeta (1890–1990) (Tanzania, Uluguru Mountains, Morogoro District, Mgeta Division)*. Thèse de doctorat, Université Paris, Paris 1. Available at: http://www.theses.fr/1993PA010728 (accessed 9 October 2024).

Coniat, D. and Raison, J.-P. (1997) Land use and spatial planning in the Upper Mgeta (Uluguru Mountains). In: Raison, J.-P. (ed.) *Essais sur les montagnes de Tanzanie*. Karthala/IFRA/Géotropiques, Paris, pp. 227–261.

Hartog, T. (2016) Agrarian diagnosis of the Upper Mgeta Region in the Uluguru Mountains (Morogoro, Tanzania): what adaptation of farmers to climate change? End of study thesis, UFR Agriculture Comparée et Développement Agricole, AgroParisTech/AFD, Paris.

Paul, J.-L. (1985) *First Progress Report on the Mgeta Zone, Report of the Agrarian Systems Survey Conducted in Nyandira in July 1985*. Franco-Tanzanian Project. University Sokoine of Agriculture, Crop Science Department/Cimade, Morogoro.

Paul, J.-L. (1988) *Farming Systems in the Upper Mgeta (Morogoro District)*. Franco-Tanzanian Horticulture Development Project. University Sokoine of Agriculture, Crop Science Department/Cimade, Morogoro.

Paul, J.-L. (2003) *Historical Anthropology of the Eastern Tanzanian Highlands: Settlement Strategies and Social Reproduction among Matrilineal People*. Karthala/IFRA, Paris.

Temple, P.H. (1972) Soil and water conservation policies in the Uluguru Mountains Tanzania, Studies of Soil Erosion and Sedimentation in Tanzania. *Geografiska Annaler Series A, Physical Geography* 54(3–4), 110–123.

IV

What Policies Should Be Promoted to Adapt to Climate Change?

_____

7 Pathways to Resilience and Adaptation: From Climate Change to Global Change

Nadège Garambois*, Hubert Cochet, and Olivier Ducourtieux
Comparative Agriculture Training & Research Unit/UMR Prodig, AgroParisTech, Palaiseau, France

---

7.1 Introduction

The various case studies presented in Chapters 1–6 (this volume) illustrate contrasting geographical contexts and degrees of exposure to climate change and increased climatic hazards (rainfall, flooding). They thus provide a range of situations rich in lessons and invite a comparison of the concrete ways in which farms adapt to climate change and hazards in the 12 areas studied, in order to draw more general conclusions. This work also reveals the unequal opportunities to adapt to climate change and hazards from one category of farmers to another, which invites a better understanding of this heterogeneity within family farming by focusing on the role that conditions of access to resources and means of production play in these processes. Finally, all this work highlights the fact that the climatic factor, its evolution, and its hazards, is not the only element that has conditioned the evolution of agricultural production systems—far from it. Farmers have had to (and still do) adapt to a whole range of global changes, of which the climate is only one aspect. The choices made in terms of public policy have also been decisive; some agricultural policies have aggravated climate change's effect, while others have been effective in increasing and securing producers' income.

7.2 Farms' Systemic Adaptation to Climate Change and Hazards

7.2.1 Manifestations of climate change and hazards on cultivated ecosystems: some typical situations

7.2.1.1 Major climatic upheavals: permanent drop in rainfall or major decline in river flooding

In addition to high rainfall variability, some rain-fed agriculture systems have experienced a reduction in the length of the rainy season, sometimes accompanied by a sharp decline in rainfall. In this case, farmers have had to make in-depth changes to their cropping and livestock systems but also adapt to strong interannual variability. This is the case in southern Tanzania (Iringa, see Chapter 2, this volume), which receives an average of 600–700 mm of rainfall, but where annual rainfall fluctuates between 500 and 1000 mm per year and where farmers report a rainy season shortened from 7 and 8 months to 6 months. In the Sahelian part of the Senegalese Peanut Basin (see Chapter 1, this volume), the shortening of the rainy season (which averaged 75 days in Bambey and 50 days in Louga over the 1968–1999 period) has for several decades been coupled with

*Corresponding author: nadege.garambois@agroparistech.fr

©2024 CAB International. *Agrarian Systems and Climate Change: Journeys of adaptation in the Global South* (eds H. Cochet, O. Ducourtieux, and N. Garambois)
DOI: 10.1079/9781800628137.0007

an average rainfall of 450–500 mm (Bambey region), and even 270–320 mm (Louga region), reflecting a drop in rainfall of up to 35% and 45%, respectively.

The reduced crop calendar window has led farmers to rethink crop type, management, and variety selection according to cycle length. The major decline in rainfall has resulted in a reduction in the biomass potentially produced per unit area, both on rainfed crops and on rangelands reserved for animal grazing, reinforced by the longer duration of bridging time. The lowering of the water table limits capillary action and tends to favor species with root systems that can explore deeper soil horizons.

In certain regions that have long had a Sahelian climate but are also close to a major river and can therefore combine rainfed crops with flood recession crops, the climatic upheaval has been twofold, as in the case of the Senegal River Delta. The drop in rainfall along the entire river course led to a major drop in water level in its downstream part and a sharp reduction in the areas where flood recession crops were possible. This decline in flood level has reduced (in volume and the concerned surface area) the supply of fertilizing elements by alluviation. It has also contributed to the strong reduction of forage resources in spaces normally submerged by the flood and usually dedicated to grazing during flood recession, as well as the collapse of fish populations and artisanal fishing activities. These upheavals have aggravated the crisis faced by rainfed crops (average rainfall of 220 mm per year over the 1968–1998 period, with almost half of the years below 100 mm, see Chapter 5, this volume).

7.2.1.2 Increased hazards: increased variability in precipitation or flood magnitude

In rainfed farming regions with climates where rainfall averages over 1000 mm per year, such as the Zambian highlands (see Chapter 2, this volume), climate data do not indicate any major changes in climate over the past 20 years, for example in rainfall volumes or temperatures, where the upward trend remains slight. In these regions, the element described by farmers as the most determining factor is rainfall's very random nature (700–1250 mm of rain depending on the year) and the increased frequency of violent rain events.

In flooded or flood recession agriculture, the cases studied sometimes show greater variability not only in rainfall but also in flooding, as in the valley of the Rufiji River and its tributaries in Tanzania (see Chapter 3, this volume), with differences of more or less 2 m in water level. Without planning to control flooding, this high variability can lead, depending on the year, either to real floods that damage crops or to farmers being forced to sow only part of the fields of land that they dedicate to flooded crops in a "normal" flood year. In addition, since the end of the 1990s, there has been a net rainfall decrease during the short rainy season and a randomness to the first rains' arrival, as well as a shortening of the long rainy season, all of which are detrimental to rainfed crops, and reminiscent of the context in the Central Peanut Basin.

Around Tonle Sap Lake in Cambodia, as in Thien Trí in the Mekong Delta of Vietnam (see Chapter 4, this volume), the scarce unflooded land is occupied by housing, orchards, and roads. Agriculture has long been based on flooded rice cultivation on these vast floodplains. In the Tonle Sap, floodplain cultivation is subject to increasing variations in the flood's onset, initial rate of water rise, flood magnitude, and then water recession rate. Farmers must deal with a single cultivated ecosystem, in which the total water control allowed by the practice of rainy season irrigated rice cultivation concerns only a part of the fields. The rest of the land is used for flooded rice cultivation, which is subject to the increasing vagaries of the flood's timing and level, which, depending on the year and the fields' position within the floodplain, can lead to overflooding or, on the contrary, to insufficient flooding, which is harmful to rice cultivation. In the Mekong Delta, flooding is dependent not only on rainfall and the river's own water rise but also on the tide's rhythm, without saline intrusion in this case. However, it commands an intraday variation in water level, adding yet another degree of complexity to managing the flood hazard. The cultivation of this particularly complex ecosystem was made possible by its progressive planning (construction of raised dikes, canals, and channels) aimed at controlling the water level at the field level in the context of flood hypervariability. This region is now subject to increasing rainfall and rising sea levels, which directly affect flooding and raise fears of an increase in the frequency of exceptional and devastating floods, whose past manifestations, although rare, remain vivid in farmers' collective memory.

7.2.2 Adaptation in the choice of species and varieties, in crop management, and by strengthening agrobiodiversity

The field studies presented here show that, in the face of climatic deterioration and increased rainfall hazards in rainfed farming or flooding in flooded farming, farmers are striving to make their cropping systems more robust by strengthening, as far as they are able, agrobiodiversity and biological regulation within their cultivated fields.

Fieldwork in the Tanzanian highlands has shown that corn is more sensitive to the delayed onset of rains than other crops and that there is a greater risk of growing solely corn. It also shows that yields from crops grown in combination are often less affected by climatic events than those of sole crops.

When the hazard of the first rains' arrival is coupled with a major climatic deterioration, farmers tend to increase the number of species and varieties of their cropping pattern that are most likely to ensure a harvest. In the Peanut Basin, the maintenance of rainfed crops grown before the drought that began in the late 1960s was made possible by the use of shorter-cycle varieties, in sync with the new rainfall conditions. Farmers are more likely to favor crops that are less sensitive to late or low rainfall. The increase in cowpea cultivation in the Peanut Basin (as a sole crop or in association with millet or peanuts) is very illustrative of these strategies: less profitable in good climatic years, but also less costly to establish than peanuts, it combines the advantages of legumes with a range of varieties that make it possible to fall back on a 45-day cowpea, known as a "rescue cowpea," in the worst climatic years, and still guarantee a minimum harvest. Some farmers have also been able to grow higher value-added rainfed crops that require low rainfall (watermelon), within the limits of the markets available to them.

In regions combining declining rainfall and the flooding hazard (Rufiji River in Tanzania), the succession of rainfed corn/flooded rice on the same field submerged for part of the year has thus evolved into a corn + rice association: a good corn harvest in the first cycle thanks to favorable rainfall can compensate for a poor rice harvest in the following cycle if the flood arrives too late, and vice versa (see Chapter 3, this volume).

This biological diversity and the greater place given to cultivated species that are likely to be more resilient to strong hazards appear to be powerful antirisk levers.

7.2.3 New crop and livestock linkages and strengthening the role of legumes in rainfed agriculture

Faced with the reduction or high variability of available fodder resources (decline in biomass on the rangelands, reduction in crop residues following the drop in yields) and the decline in livestock, which are due in part to climatic deterioration, farmers have been able to make in-depth changes to their livestock production systems (animal and breeding types, feeding methods, etc.).

Thus, in the Sahelian Peanut Basin, after a strong decapitalization of livestock at the height of the drought period in the 1970s and 1980s, farmers reconstituted smaller herds, leaving more room for small ruminants, and carried out profound labor intensification in the feeding of their livestock and in the transfer of organic manure to cultivated fields, thereby increasing the efficiency of these transfers and the closeness of the links between crops and livestock. Animals are now most often kept in stalls, fed crop residues (millet straw and bran, peanut and cowpea haulm, or even cowpea forage fields strictly dedicated to feeding livestock) collected directly from the cultivated fields by each family, which then transfers dry livestock dung to its fields. The pruning of trees in the wooded area, particularly *Acacia albida*, supplements feed during periods of low fodder (hunger gap). However, this practice poses some difficulties when it becomes excessive and threatens the species' sustainability, which is essential to the agrarian system's balance (atmospheric nitrogen fixation and fertility transfers, see Chapter 1, this volume).

In the Senegal River Delta, where drought has prompted the government to accelerate hydraulic planning and the development of irrigated rice cultivation, farmers who used to herd their animals on rangeland (land that is unflooded in the rainy season and flood recession areas in the dry season) now feed their animals partly in stalls thanks to the collection of rice straw, which has not, however, been enough to prevent a sharp decline in the number of cattle.

Despite the intensification of these transfers, the decline in livestock numbers on a regional scale partly explains the maintenance, and even the increase, of the place of legumes in rainfed crop rotations when conditions of access to chemical fertilizers are more difficult (Sahelian Peanut Basin, see Chapter 1, this volume, and Tanzanian highlands, see Chapter 2, this volume).

7.2.4   In the face of hazards, offseason crops (irrigated or lowland) offer a decisive opportunity to reduce vulnerability

In rainfed regions faced with unfavorable climatic conditions (shorter rain duration, sometimes reinforced by a decline in rainfall), farmers who could afford it turned to the use of lowlands or the planning of small, irrigated areas, which could be used in the dry season and were less subject to rainfall vagaries. The increased use of these areas also had the advantage of using family labor at a time when the work calendar was not saturated.

In the Tanzanian highlands, the planning of small, irrigated perimeters equipped with gravity irrigation systems has also made it possible to secure corn cultivation in the rainy season and grow offseason vegetable crops. This cultivation of the lowlands (all year-round or only in the dry season, depending on their flooding conditions in the rainy season), although also dedicated to corn cultivation, does however allow access to new, much more profitable production that is harvested and sold fresh, such as corn cobs. In North Zambia, the use of small, irrigated fields for sale or for self-consumption, in addition to rainfed crops, also significantly reduces families' vulnerability (see Chapter 2, this volume).

The development of offseason market gardening in the Uluguru Mountains (Tanzania),

thanks to farmers' continuous planning of gravity irrigation networks on the hillsides, offers another spectacular example of the gradual filling of the farmers' work calendar and the diversification of their production and sources of income (see Chapter 6, this volume).

Finally, the hydraulic planning of the Senegal River Valley, initiated in the 1950s in the north of the delta even before the droughts of the 1970s and 1980s, also provided farmers who had access to irrigated areas with the opportunity to diversify and at least partially secure their income. The hydraulic planning of areas dedicated to flood recession crops for the benefit of flooded and then irrigated rice cultivation, as well as offseason market gardening, accelerated significantly from the 1970s to 1980s. Indeed, larger-scale planning was then aimed at rapidly securing cultivation conditions through total water control, completed with the construction of the upstream and downstream Manantali and Diama dams, which made it possible to control the river's flooding and prevent saline rises from its outlet. Most of the fields in these areas are now supplied by gravity irrigation, from secondary canals supplied by pumping stations, or directly from individual or collective motor pumps installed on the main canals. These operations have been supplemented by the planning of irrigated perimeters on unflooded land, primarily dedicated to market garden crops (see Chapter 5, this volume).

7.2.5   Adapting to lack of water control in flood-prone areas

We have seen (see Chapter 3, this volume) how farmers in the floodplain of the lower Rufiji River Valley (Tanzania) had gradually shifted the sowing of corn, which precedes rice, due to the rains' later arrival, to the point where the old corn/rice succession has become a succession/crop combination, with the two crop cycles overlapping to a considerable extent. The rice cycle then offers a chance to catch up with those who planted it on time. Conversely, a good first-season corn crop (which implies an early rainfall onset and sufficient volumes) can compensate for an unfavorable rice cycle due to floods' late arrival or their receding too early. Farmers also take advantage of intraplot heterogeneity, as the fields are not flat—far from it. In this case, seedlings are staggered and spread out, from the lower parts of the fields to their "tops," this practice being an additional means of harvesting, even in the most unfavorable rainfall and flood conditions. In addition, when flooding too early and too quickly destroys rice seedlings, farmers have begun, at great labor cost, to pull up some of the rice plants in the less damaged parts of the fields (slightly elevated) to do actual transplanting in the lower parts of the fields where the heavy flooding has destroyed the first seedlings.

In the Tonle Sap Basin in Cambodia, the varietal diversity and adaptation of practices to various flood conditions is a source of resilience and risk reduction. Transplanting, which is more widely practiced here than in the Rufiji Valley (Tanzania), also plays a major adaptive role by allowing the transplanting date to be closely adapted to the level of water flooding the fields. The last resort, if the climatic hazard has affected the nursery itself, may then be direct drilling on the highest and therefore least flooded land (see Chapter 4, this volume).

The hypervariable and intraday nature of the water level in the Mekong Valley (Vietnam), due to the three elements of rainfall, river flooding, and tide, has progressively led to the search for total water control and the development of irrigated rice cultivation to control the hazard and make maximum use of the productive potential of the fertile combination of abundant water and a floodplain silted up every year. Since the 1970s, access to motorized pumps to facilitate drainage during floods and irrigation in the dry season, non-photoperiodic varieties of rice with a short cycle and high yield potential, and subsidized inputs have made it possible to conduct two rice cycles per year on part of the floodplain and increase the yield per hectare. This technical Green Revolution also owes its results to the accompanying agrarian reform. The elimination of land rent and uncertainty of sustainable access to land has given three quarters of peasants the security and means to invest in their family farms. The completion of diking and canal opening works at the end of the 1970s and the gradual generalized use of power tillers during the 1980s–1990s made it possible to set three rice cycles per year in most rice fields. A third major series of planning carried out and financed by the State in the 2000s, aimed at raising and concreting the dikes again, made it possible to secure the development of orchards on the upper lands (crops with higher value added than rice, see Chapter 4, this volume).

7.2.6   Complementary use of diverse "terroirs" as an antirandomization technique

In flooded agriculture (Rufiji River Valley in Tanzania, see Chapter 3, this volume) as well as in flood recession agriculture (Senegal River Valley in the 1950s, see Chapter 5, this volume), staggered sowing aligned with the advancing or receding flood and the use of staggered story fields, in variable conditions subject to flooding or its recession contribute to limiting risks. Here, what reduces risk is the use of fields that are heterogeneously subject to flooding. The combination of rainfed crops and flooded or flood recession crops, grown on different terroirs subject to different hazards (rainfall or flooding), offers greater robustness to agricultural production systems, provided that these differentiated and complementary areas are accessible. In the Tonle Sap Basin in Cambodia, it is possible to grow rice at various topographic and agroecological levels, which are located in different conditions with regard to flooding and flood recession. The economic vulnerability of the farm households depends on their capacity to combine the access to multiple levels (see Chapter 4, this volume).

In the Uluguru Mountains of southern Tanzania, there is access to several ecological stories that, along with irrigation, allows families to combine as many crops and crop varieties as possible by adjusting crop cycle positioning to environmental conditions and the availability of family labor. This diversity thus makes it possible, while limiting risks, to efficiently employ the family labor force throughout the year and thus increase overall labor productivity and farm income (see Chapter 6, this volume).

Thus, access to different ecosystems and the combination of different cropping systems and, where appropriate, different livestock systems (including fishing), makes it possible to limit risks and take maximum advantage of the complementarities provided by the range of activities available to farmers. The combination of the exploitation of different spaces by the same family is a guarantee of limited vulnerability, and the households with the possibility to exploit all these landscape "facets" generally fare better than others. The families in the best position, those with the highest incomes and that are most equipped to face the different types of hazards to which they are exposed, are always those that can combine, in different spaces, the greatest possible number of complementary crops and activities. This is provided that they have the necessary means of production, an appropriate work capacity, and access to markets under acceptable conditions.

In contrast, history teaches us how particularly coercive experiments involving the forced regrouping of rural households and compulsory work within collective fields, such as the villagization carried out in Tanzania in the 1970s, run counter to these adaptation strategies. This has led to longer daily transport times, making access to the most remote ecosystems of the new village particularly difficult, complicating the management of animals, and leading to their excess mortality. Despite the support mechanisms provided to farmers for corn production, this authoritarian policy has made it much more difficult for farmers to adapt their production systems to the climatic deterioration that has occurred in parallel.

7.2.7   Long-standing adaptation levers, based on farmers' expertise

The case studies presented in this book reveal that the range of antirisk techniques used by farmers under hyperrandom climatic conditions often reflects a long-standing capacity for adaptation, closely linked to the farmers' detailed knowledge of their environment and the gradual development of expertise.

The case of regions that were already Sahelian in the first half of the 20th century (northern Senegalese Peanut Basin, Senegal River Delta) also provides an eloquent example of this expertise. At that time, delta agriculture combined a whole range of productions through a complex and non-random combination of rainfed and flood recession crops, mixed livestock farming and fishing, mobilizing the differentiated but complementary use of different terroirs and playing on herd mobility (transhumance) to better adapt to variable grass growth on the rangelands. Low population density, multiplication of ways to store different food products (cereals, meat, fish) at different scales (granaries per household, extended family, village), bartering between producers with varying degrees of specialization (including strict pastoralists), and complementary extra-agricultural activities were all factors that contributed to the robustness of this agrarian system and limited the vulnerability of different categories of producers (see Chapter 5, this volume).

In the Peanut Basin in the late 1960s, the southward shift of isohyets placed the center of the basin in rainfall conditions close to those of the north of the basin before the drought. Farmers in the center were thus able to mobilize the species and varieties that had previously been selected by farmers located further north. The assistance of agronomic research, which made it possible to select and rapidly disseminate millet, cowpea, and peanut varieties with particularly short cycles, also facilitated the adaptation of agriculture, particularly in the northern part of the Peanut Basin, where rainfall had fallen to a level barely maintaining strictly rainfed agriculture (see Chapter 1, this volume).

In the Kilombero and Rufiji floodplain valleys in southern Tanzania, families' survival relies on a detailed knowledge of the floodplain's microheterogeneities and the skillful exploitation of this very complex and inartificial environment. The example of corn + rice intercropping on the sandbanks and floodplain of the lower Rufiji Valley provides an eloquent example of this fine and evolving adaptation to the vagaries of rainfall and flooding. Here, associated cropping is not only a "strategy" for anticipating and reducing risk, which makes it possible to "not put all one's eggs in the same basket"; it is also a means of adapting and steering *in itinere* by taking what has happened at the beginning of the rainy season (date of the first significant rainfall, spacing of the first rains, rainfall volumes, etc.) into account.

7.3 Access to Resources and Means of Production, a Factor in Unequal Adaptation, and Robustness of Family Production Systems

7.3.1 Specific responses according to context and families' access to resources

The results of the case studies presented in the previous chapters show that even families in the most difficult situations, those with the most limited and least diversified access to land, livestock (and therefore to organic manure), and means of production, are nevertheless trying to adapt to the changes that affect them. For example, faced with the end of subsidies for peanut seeds and inputs, most families in the Senegalese Peanut Basin equipped with at least one seed drill (and sometimes a cultivator) have chosen to rebalance the share of their crops between peanuts and cowpeas and preserve a minimum on-farm consumption (millet cultivation) before considering sales crops if external income is not sufficiently secure. In addition, since these families can only ensure very low recycling of nutrients in their fields (in both organic and mineral forms), legumes' importance in their cropping pattern is increased. In some cases, the poorest families, with the smallest land areas and the least equipment, have become relatively specialized in the production of certain seed legumes that are largely intended for sale. In this case, these families engage in calorie substitution, with the sale of more expensive calories (seed legumes) being used to purchase basic cereals. In the northern part of the Peanut Basin, this is mainly cowpeas. Indeed, peanut cultivation is highly vulnerable to rainfall, and therefore much too risky for these families.

Similar strategies can be observed in North Zambia, where not all farmers have the possibility of combining rainfed and irrigated crops. Even without access to irrigation, families manage to play on complementary access to different terroirs in order to generate an income above the poverty line, while favoring family on-farm consumption as an antirisk technique.

However, access to varied terroirs with complementary potential and different exposure to hazards remains key to reduced vulnerability and enhanced adaptive capacity. Thus, farmers

in the Tonle Sap are still trying to take advantage of the basin's microheterogeneities to combine different rice-growing systems throughout the toposequence and flooding and flood recession timeframe (see Chapter 4, this volume). In the Rufiji River Valley, this diversification is based above all on the association of fishing and farming, the minimum basis for security being carrying out fishing and farming activities in parallel, which presupposes a balanced presence of adult men and women in the household. This is based on access to different landscape "facets" and their joint exploitation; exploitation that is more successive than simultaneous, due to the absolute need to monitor fields against wildlife intrusion. Families are unequally exposed to hazards. For instance, if they only have access to fields that are subject to flooding but located on sandy soil, they are more vulnerable in the event of a low flood (insufficient residual moisture). They also can have access to unflooded fields dedicated to rainfed agriculture (subject to the rainfall hazard). Finally, they can combine access to these different spaces and access the most fertile fields on the floodplain (see Chapter 3, this volume).

In agrarian systems where rainfed agriculture dominates, access to a few irrigable fields appears to be a powerful means of increasing this diversity and farm resilience. In the Tanzanian highlands, the rains' earlier ending considerably increases interest in areas (lowlands or small irrigated areas) where it is possible to grow two corn crop cycles per year. Here, access to irrigation is an essential factor in increasing labor productivity (more capital-intensive sales crops, but much more remunerative per surface area unit) and robustness in the face of less favorable rainfall conditions in certain years.

Despite the sometimes difficult—and uncertain—marketing conditions for offseason vegetable crops, access to irrigation, even in small areas, can be a determining factor in the income levels recorded by families but remains very unevenly accessible. This is the case for gravity irrigation in North Zambia on artisanal perimeters, which does not require heavy investment but raises the question of access to areas where this irrigation is possible. In the Senegalese Peanut Basin, in the absence of surface water, the planning of irrigated market gardening areas relies on the construction of wells and on investment capacities that, apart from NGO projects, are only available to the few families with an expatriate member in Europe. The practice of market gardening also assumes that these farmers have sufficient organic manure to ensure the recycling of nutrients and maintenance of the fields' organic matter content, even though these crops are demanding in terms of fertility and produce little, if any, crop residue for livestock.

In central Zambia and the Senegal River Delta, where real irrigated perimeters have been developed for a longer period of time, mainly for rice production and market gardening (onions, industrial tomatoes, etc.), the conditions of access to fields irrigated by gravity or a motor pump placed along the primary canals or main rivers, remain very unequal. With very limited access to these perimeters and only being equipped with manual tools (watering cans), most families in the Mkushi region (Zambia) are only able to grow offseason vegetable crops in very small areas (less than 0.1 ha); slightly better-off families have been able to set up their own small furrow irrigated perimeters (0.6–0.7 ha). The wealthiest families are equipped with a motor pump and have 1–2 ha of market gardening crops, for which they use very low-paid hired labor.

This last example illustrates that access to diversified land with complementary potential must also be accompanied by sufficient access to means of production: irrigation equipment, draft animals and adapted tools, storage equipment, and post-harvest processing. In the north and center of the Senegalese Peanut Basin, there is also access to draft animal power equipment, land, and livestock that has for several decades determined the area that can be planted and the respective place of different crops (millet, peanuts, cowpeas) within the cropping pattern. The climatic deterioration and the narrowing of the calendar window have made access to mechanized sowing and weeding even more crucial, especially for peanuts, the only crop that cannot be directly sown in dry soil (at the risk of losing a very expensive seed in the event of delayed rains).

Finally, in the different regions studied, the best-off families often specialize in food crops intended for sale in urban markets (large towns, cities), which are generally riskier and more capital intensive: peanuts in the Senegalese Peanut Basin; industrial tomatoes, onions, and chili peppers in the lower Senegal River Delta; corn

and market gardening in Zambia and South Tanzania; irrigated rice in the Tonle Sap (Cambodia); irrigated rice and fruit in the Mekong Delta (Vietnam). This relative specialization does not prevent priority being given to family on-farm consumption, which is facilitated by abundant cash flow, access to the most suitable fields, and equipment and facilities to limit risks, as in the case of the largest farms in South Tanzania (Kiponzelo), which are largely dedicated to market garden production and equipped with mechanical or petrol pumps. In the north of the Senegalese Peanut Basin, the wealthiest families' advanced specialization in leguminous seeds (especially peanuts) for sale is clearly based on securing the family's food supply through external migratory income.

7.3.2 Cumulative inequalities at the root of large differences in labor productivity, generated surpluses, and adaptive capacity

The social inequalities between farmers encountered in all the regions studied concern (i) the surface areas and types of fields to which farmers have access (access to the most potentially productive ecosystems); (ii) their level of equipment and capital enabling them to develop these areas (planning and hydraulic equipment); and (iii) their capacity to renew the fertility of cultivated ecosystems. The latter case is based mainly on the duration of regrowth from slash-and-burn or clear-and-burn systems, density of the leguminous tree stand, number of livestock and potential for transferring manure (cart or backpack), use of synthetic fertilizers and soil improvement products, and so on. Different categories of producers thus do not have equal means to deploy adaptive capital intensification strategies (tools, irrigation, livestock, seeds, etc.).

These inequalities are cumulative by generating larger surpluses and sometimes being able to rely on solid family financial contributions (urban, expatriate), the richest families are also in the best position to invest in equipment that will allow them to add value to their agricultural production by going beyond the "edge of the field." This consists of artisanal processing made possible by mechanized equipment that allows them to process large volumes and provide services (peanut processing in central and northern Senegal), as well as the acquisition of motorized means of transport to sell their market gardening and cereal production directly to the final consumer while developing purchase resale from neighbors without the means to access urban markets (as in the case of the Iringa region in southern Tanzania).

With very small surpluses, the poorest farming families appear to be particularly vulnerable to changing climatic conditions and to the variability of agricultural prices, both because they can only market small surpluses, and they are also often forced to specialize in "luxury" agricultural production, the demand for which follows growing urbanization. They also specialize in temporary salaried farming workforce, in order to buy low-cost calories in return. Unable to feed their families in any other way, these farmers appear particularly exposed to climatic hazards as well as to agricultural and food prices. A bad year in terms of weather can plunge them into a longer hunger gap and force them to fall back on salaried work (sometimes far from their homes) to ensure the family's survival.

In the central Peanut Basin (see Chapter 1, this volume), where families still live mainly from their agricultural activity, income differences per worker and per year vary by a factor of five, between poor young households with little and underequipped land, and better-off families able to market surplus millet and peanuts. In the lower Rufiji River Valley, and in a context where the differences between families in terms of equipment are insignificant, the differences in income are nevertheless one to four, between households whose fields are located in the areas most vulnerable to flooding and those with access to all the terroir and the most fertile flooded fields.

In the rainfed highland regions of Zambia and Tanzania (see Chapter 2, this volume), farm incomes are only a few hundred euros per worker per year, similar to the majority of farmers on the continent who are engaged in predominantly rainfed and manual farming. However, those who are able to generate better and more regular incomes and protect their families from all kinds of hazards are almost always those who implement diversified production systems, relying on a wide range of accessible ecosystems, intercropping, and small-scale irrigation. In the steep hillsides of the Uluguru Mountains, the key to success, along with access to water and proximity to the market (on the roadside), is the altitudinal spread, when one has access to the different ecological stories. In addition to inequalities of access resulting from the arrival order and conditions of different lineages in the region during the pioneer phase, there are inequalities inherent in the differentiation processes that have been quite pronounced since the second half of the 20th century, and that are especially linked to conditions of access to irrigation, market gardening, and the market. Moreover, with the rise of cash crops, but also the increase in land pressure and difficulties in accessing land, the gap has widened between families with large farms that call on a workforce from outside the family, and households that do not have enough land to meet their needs and are forced to sell part of their labor force to ensure survival (see Chapter 3, this volume).

In Cambodia, on the banks of the Tonle Sap, the best-off families deploy very capital-intensive adaptation techniques. The acquisition of a multi-purpose tiller allows them to gain considerably in physical labor productivity while reducing the risks associated with planting rice at the beginning of the rainy season. The speed of motorized plowing, combined with the possibility of dry soil tilling, makes it possible to widen the calendar window by being free from soil becoming wet from the first rains, which is essential for mechanized (draft animal power) or strictly manual tilling. At the same time, those families who have fields close to the canals replace part of their rainy season flooded rice cultivation—subject to the flooding hazard—with irrigated dry-season rice cultivation. The development of irrigated rice cultivation creates more value added per unit area and per labor unit, provided that conditions of access to capital and water are met, but it also contributes to a considerable widening of the income gap between producers. In Vietnam, families with land and equipment have higher income per worker than that earned from low-skilled urban jobs; on the other hand, the income of families "forgotten" by the agrarian reforms is much lower, fueling a continuous rural exodus over the past half century.

7.3.3 The limits of technical package dissemination: dependence, increased risk, inequalities of access, and possible misappropriation

In parallel with these changes in climatic conditions, farmers have often been confronted with standardized and imposed technical advisory frameworks: more informational for peanuts in the Peanut Basin, strongly incentive based in Zambia for corn (sole crop cultivation, row seeding, and monoculture promoted by relying on "model" control farmers in villages), and tightly supervised for rice production in the Senegal River Delta, with farmers in the 1960s and 1970s then reduced to the role of mere implementers in the large perimeters managed by the Société d'Aménagement et d'Exploitation du Delta du Fleuve Sénégal (SAED).

Today, the purchase of fertilizer in the "large perimeters" of the Senegal River Delta, made by each producer in the Hydraulic Unions in a fixed quantity per hectare, does not equate to the actual distribution and use of these fertilizers within each Union. The poorest farmers, who do not have the cash to purchase these subsidized fertilizers, sometimes informally sell their right to purchase them. They make arrangements with a wealthier farmer, who lends them the full amount covering the purchase of the subsidized fertilizer to which each farmer in the Union is entitled, in proportion to his or her cultivated area in the perimeter. The lender then recovers most of the fertilizer for use on rice fields in his or her private perimeters that are not eligible for subsidized fertilizer (see Chapter 5, this volume).

From the 1950s until the Structural Adjustment Programs (SAPs), the dissemination of these subsidized technical packages was based, on the one hand, on the control and support of the prices of the agricultural products in which farmers were encouraged to specialize, and, on the other hand, policies that facilitated access to agricultural equipment (in the 1960s, draft animal vehicles in the Senegalese Peanut Basin). During this period, this support of the agricultural sector, for example in Zambia (corn) and Senegal (peanuts), contributed to a rise in physical labor productivity and a strong increase in production and producers' income. This secure agricultural policy framework contributed to a certain specialization of production systems and made them less autonomous in terms of fertility restitution and therefore more dependent on mineral fertilization.

This dependence, already tangible in Zambia before the end of the 1980s and directly linked to the Zambian State's fluctuating capacity to ensure these production conditions, became glaring not only in Zambia but also in Senegal with the introduction of SAPs and the abrupt end of these schemes in the 1980s. With the end of input subsidies and price guarantees for producers, the use of these technical packages became inaccessible and much too risky for most. Without sufficient parallel support for agricultural prices, only the richest farming families, whose food is ensured by sufficient cereal production or purchases thanks to money from emigrant relatives, have the means to buy "improved" seeds and fertilizers and maintain (or even reinforce) this specialization, by improving their products' value (illegal transformation of downgraded peanuts into cattle oilseed cake in the Senegalese Peanut Basin, for example). In the Peanut Basin, as in the Tanzanian highlands, only these families have the means to engage in this highly profitable production, but it is costly to set up and risky in years when climatic conditions are less favorable.

The spread of the System of Rice Intensification (SRI) in the Kilombero Valley (southern Tanzania) provides another example of a technical package that is potentially effective but carries an increased risk for farmers. It includes selected short-straw varieties, row seeding, and mineral fertilization. Although yields are much higher (up to 7 t/ha of paddy) when conditions are optimal, the adoption of the SRI technical package only concerns a very limited area on a given farm, because the overall budget for a year of SRI cultivation is so high that a poor harvest would be catastrophic for households if all the available area were managed in this way. By dedicating only a small portion of the land to SRI and the other portion to the "traditional" system, it is possible to hope for a better harvest in a normal year, without incurring excessive costs. This is a way to increase income while limiting the risk of climatic hazards.

On the banks of the Tonle Sap, the increase in climatic hazards leads to a likely delayed start to the rainy season, during which partial water control limits the possibilities of developing SRI in the floodplain. This is because (i) transplanting very young (5–10 days) and therefore short seedlings increases the risks of loss in the event of rapid and early flooding and (ii) using this technique implies total water control to alternate the field's flooding and drying phases, which is only possible for farms that have access to irrigated rice fields in the dry season. Moreover, the potential increase in yield per hectare is dependent on an increased amount of work at a time of year when labor is already scarce, and an increased risk of flooding (see Chapter 4, this volume).

Here, as in many regions of the world, the first objective put forward by public authorities and the companies and projects that accompany them is always to increase yields, presented as an end in itself for producers. However, for a family farm, the objective is not necessarily to have the best yields, but to generate sufficient income to support the household, limit the risks of obtaining an income that is incompatible with the family's survival, and potentially invest. In the case of the SRI program, we note that the management sequence that has been popularized and is supposed to increase resilience to climate change results in an increase in the risk incurred by farmers, and that this increase is the main limit to these practices' dissemination.

Although voluntarist policies to promote a particular crop and efforts to improve varieties and disseminate fertilizers have sometimes borne fruit, when funding enabled effective subsidies for these inputs and when they were available in a timely manner, it is clear that the "corn technical package" or the "peanut technical package" alone have not brought about a lasting improvement in rural people's conditions. Vietnam and its voluntarist policy of disseminating the Green Revolution, which was accompanied by the long-term management and financing of hydraulic planning, agrarian reforms, input and equipment subsidies, rural credit, and guaranteed rice prices, is a remarkable example in this respect (see Chapter 4, this volume).

In any case, in the regions studied here, most development programs, now often carried out by NGOs and sometimes producers' organizations, still rarely consider farms in their complexity, as a combination of productions, possibly competing against each other for the allocation of resources, but also complementary to one another in terms of work, cash flow, and food calendars, and in the use of organic manure, fodder, and crop residues. Programs also rarely take into consideration the diverse production conditions in which producers may be placed within the same region. Many programs thus remain focused on a single production (corn in Tanzania, rice in the Senegal River Delta, cowpeas that succeed peanuts in the Peanut Basin) and provide conditions that are unsuitable for many families (minimum surface area required by the program that covers most of their farmland, specialization that is too risky, availability of capital and equipment that is too limited, etc.).

7.4 Beyond Climate Change, Farmers' Adaptation to Global Changes

7.4.1 Farmers' adaptation to a combination of changes

While in each of the regions studied, climatic deterioration or the increased frequency of its hazards has influenced farmers' practices, their respective production choices are also directly linked to all global changes in addition to the climatic ones that have taken place in parallel: demographic growth and a downward trend in surface area per worker, falling and increasingly volatile agricultural prices, development of domestic transport and growing integration into market exchanges, reduction of tariff barriers and competition with moto-mechanized agriculture with a high level of physical labor productivity, the emergence of new local demands for food products due to growing urbanization, and so on.

Faced with changes in the global economic context, international trade rules, and world prices of agricultural commodities and inputs, national agricultural policy frameworks have undergone profound upheavals, marked by the collapse of state agencies that provided subsidized inputs and equipment and guaranteed the sale of production at a price significantly higher than the world market price. This break in the support system also led farmers to adapt to the drop in domestic prices for certain crops that had previously been heavily supported (peanuts in Senegal from the mid-1980s, corn in Zambia and Tanzania from the late 1980s) and often to the fact that they could no longer access synthetic fertilizers and seeds with higher yield potential.

The main developments noted as factors of adaptation to changing climatic conditions are therefore also largely due to systemic adaptation to this new and uncertain context in many areas. Labor intensification in crop and livestock production, the broadening of the range of productions and the saturation of the work calendar through offseason crops, and the increased use of legumes for the domestic market in cropping patterns are all strategies that aim to offset the decrease in the available surface area per worker, drop in cattle numbers, and much more costly access to mineral fertilizer. By taking advantage of a wider range of crops, farmers also have a better chance of reducing their vulnerability to variable agricultural prices, which do not affect all crops to the same extent in a given year.

The development of higher value-added products is part of the same logic. These products are often perishable (market gardening) or sought-after by urban dwellers for their typicality (fresh corn cobs, peanut oil, local rice, bissap, etc.), for which family farming with low labor productivity still has real comparative advantages, despite the generalized lowering of tariff barriers. Preserving a minimum level of food self-sufficiency whenever families have the means to do so is again an antirisk technique in the face of climatic hazards, but also in the face of the high variability of imported grain prices, which are most often indexed to those of the highly volatile world market.

7.4.2   Strategies for expanding farm household income sources: the rise of processing and off-farm activities

In addition to adapting their farming practices to this new climatic and global context (population growth, variations in agricultural prices, conditions of access to inputs and equipment, urbanization and demand on domestic markets, particularly in urban areas, etc.), many farm households are increasing their agricultural product processing activities (or fishing in regions where this is possible), to varying extents according to their means, and are also seeking to diversify their income sources through off-farm activities. This may involve sending part of the family to work in the city during low agricultural activity periods, but also sometimes working on other farms as salaried agricultural employees throughout the year, when families do not have the means to ensure their survival with agricultural activity alone.

In the northern part of the Senegalese Peanut Basin, the severe drought experienced by this region in the 1970s and 1980s, reinforced by the drop in producer peanut prices in the mid-1980s, amplified the labor migration that was already taking place in this part of the country. Faced with the demand for low-skilled labor from European countries, particularly in the construction industry, these migrations took the form of real expatriations for members of families able to finance the trip for one of their number. Today, these families have substantial external income, which contributes to securing their food supply and explains their specialization in agricultural production for sale, which requires more capital.

On the banks of the Tonle Sap (Cambodia), some of the youngest workers can move to Thailand to work in very low-skilled jobs (agricultural day laborer in rubber plantations, laborer or employee in the industrial and service sectors). Their precarious status—politically maintained in Thailand to contain the cost and ensure the docility of this immigrant labor force—makes their external income particularly dependent on the Thai economic situation. The living conditions of young rural women migrating to the Cambodian capital Phnom Penh to work in the textile industry are hardly more enviable (see Chapter 4, this volume).

7.4.3   Collective adaptation and resilience, more family based but sometimes also village based

Several of the case studies in this book show that family solidarity has played and continues to play a central role in adapting to and reducing family vulnerability. The intensity of family ties between the countryside and the city, and even with expatriates, as in the northern part of the Senegalese Peanut Basin, is reflected in transfers of money, agricultural products, and labor exchanges. Activities in the city are a source of supplementary income, where agriculture can represent both an investment activity and a fall-back solution. The Wolof and Peul societies of Senegal, with their highly hierarchical kinship structure, have helped perpetuate the important role of the firstborn over their younger siblings present in the village and in the management of extended family resources (food stocks, alloca-tion of labor, centralization and allocation of external income and money from migration).

The growing integration of market ex-changes, increased proportion of purchases dedi-cated to the family diet, and demographic growth seem to have reduced the weight of certain exchanges and collective antirisk techniques (decline in barter and village granaries). However, these techniques seem to have been reinvented around new practices based on collective organ-ization. In the absence of a management structure such as the SAED, which is reserved for the large perimeters of the Senegal River Delta, the establishment of operational village irrigated perimeters in Zambia, the highlands of southern Tanzania, and the Senegal River Delta has been based on sufficiently effective social water man-agement at the village level and the development of usage rules. In Zambia, these rules concern both canal maintenance (annual collective canal clean-ing subject to a fine in the event of non-partici-pation) and water distribution (monitoring water availability downstream and limiting volumes taken upstream at the water intakes, taking water freely upstream from the network using simple watering cans).

7.4.4 Voluntarist agricultural policies to better resist hazards?

The case studies presented in the previous chap-ters have shown the effectiveness of agricultural policies in terms of adaptation when they have the means to integrate all of the following:

- equitable access to the means of production
- dissemination of technical levers to increase yields per hectare for the greatest number of producers

- sufficiently stable and remunerative agricul-tural price levels
- if necessary, bearing the costs (project management and financing) of planning (e.g. hydraulic), up to the field scale
- an endogenous growth strategy, focusing initially on food sovereignty supported by family farming

The comparison between Vietnam's rice development policy and those of Zambia and Tanzania for corn, or Senegal for peanuts and rice, is quite enlightening in this respect. Although Vietnamese policy has not prevented real social differentiation in family farming in the Vietnam-ese deltas, the meteoric rise in rice production in Vietnam today is still driven by family farming and owes much to the very high labor intensifi-cation provided by farmers in the deltas (Mekong and Red River). The reasons for this success are to be found in Vietnam's voluntarist and re-assuring policy, combining (i) financing of major hydraulic planning down to the farm manage-ment scale (field); (ii) farmer autonomy; (iii) agrar-ian reforms to promote equity in access to planned land; (iv) support for access to equipment and inputs (facilitated by the absence of import taxes); and (v) guaranteed rice prices for producers and fixing the price of rice for consumption on the domestic market through tight border trade con-trol and a monopoly on rice exports. What is also striking is the continuity of this system for half a century, despite wars and drastic changes in political regimes: priority to family farming (agrar-ian reforms and farm size control), support for producers (inputs, prices), fairly tight control of the rice market adapted to Vietnam's current export position, continuation of large-scale planning with a perspective on developing other crops with higher value added, and intended for export (see Chapter 4, this volume).

Agricultural policies in the other countries studied here rarely show this level of complete-ness and never over such a long period of time. The Tanzanian and Zambian agricultural pol-icies implemented in the 1950s and 1980s in favor of corn production for the domestic market (guaranteed prices, easier access to inputs, etc.) and the Senegalese peanut policy implemented during the same period (guaranteed prices, easier access to inputs, and equipment), but for export purposes, came to an end with the Structural

Adjustment Plans. In Zambia, a public support system was restored in the early 1990s, based on an input subsidy program, guaranteed corn price, and the sale of production through the Food Reserve Agency. However, this system suffers from irregular and late supplies of subsidized fertilizer in the villages.

The policy of developing irrigated rice cultivation initiated in the Senegal Delta in the 1950s quickly came up against excessively centralized management by the *Société d'Aménagement et d'Exploitation des Terres du delta et de la vallée du Sénégal* (Senegal River Delta land development and exploitation company), or SAED, and disproportionate investment and management costs in relation to the yields and value added obtained. All of this planning was only made possible, both before and after the SAED's disengagement in the 1980s, by the assistance of foreign donors. The continuation of this planning and maintenance of attractive conditions of access to credit and equipment for producers, as well as a local rice market that is still slightly protected, explain the overall expansion of private perimeters since the 1980s. This new development has benefited from the bullish global rice market since the mid-2000s. The talk of national rice self-sufficiency is no longer accompanied by a desire to ensure increased production through equitable development among producers, as it was in the 1960s and 1970s. It now leads to fears of increased social inequalities in the Senegal River Delta, where the rise of new forms of agriculture (patronal and especially capitalist) in direct competition with family farming, is already well under way (see Chapter 5, this volume).

7.4.5 The emergence of capitalist forms of production: a threat to family farming

The rapid and massive circulation of capital, recent global financial crises, and upward trends of certain agricultural prices, which are highly fluctuating and encourage speculation, have brought to light new risks for family farms in regions where there is a combination of (i) the existence of land designated as underexploited by public authorities within village boundaries, (ii) land allocation rules that do not sufficiently protect village families, and (iii) the possibility of developing highly profitable production through

high investments (facilities, equipment) in a context of low-cost access to land and labor.

In several of the regions studied, farmers have been confronted, in some cases since the 1980s and in a way that seems to have increased over the last 10 years, with the development of capitalist-type forms of agriculture. Their economic rationale (increasing labor productivity by cultivating vast areas and using high-capacity equipment) and their very high availability of capital place them in extremely favorable productive conditions, and in a position to exert deleterious competition with regard to the surrounding family farms' weak resources. These forms of agriculture are very demanding in terms of land, and today, where conditions of access to land and water have allowed them to be established, they constitute a serious threat to the family farms with which they coexist. Their establishment is often an additional source of uncertainty and risk for rural people, as jobs offered by investors are generally few and precarious.

The evolution of agriculture in the Senegal River Delta, particularly in its southern part, is emblematic of these phenomena. The development of irrigation, which came later than in the Upper Delta and was more widely implemented after the SAED's disengagement in the early 1980s, has relied more on donor financing (primary and secondary networks), private planning (tertiary network), and the use of expensive motor pumps that are only accessible to investors from outside the village. The total control of flooding from the end of the 1990s in all the main rivers of the region (Senegal River and its main distributaries), the evolution of land management in the delta marked by decentralization (land transfer in the "terroir zone" in 1987 and issuance of property rights on land devolved to communes and villages), and the parallel disengagement of the SAED benefiting the planning of private irrigated areas all marked the beginning of a race for land with the irrigation possibilities offered by the various waves of primary and secondary planning (see Chapter 5, this volume).

While corporate farming is virtually absent from the Mekong Delta, its expansion has been under way in Cambodia since the 2000s. After the first major project (1 million hectares) was aborted on the banks of the Tonle Sap—which was supposed to involve the Kuwaiti State as an investor—it is now national entrepreneurs, mainly from Phnom Penh, who are investing in more limited, and therefore more discreet, projects. The productivity of these large, irrigated rice farms is lower per unit area than that of family farms, while depriving the latter of easily irrigated areas that would be very useful for securing their production and income (see Chapter 4, this volume).

In Mkushi District (Central Zambia), the post-1950s expansion of large-scale irrigation schemes with diversion canals and gravity irrigation followed by direct pumping into the main river has allowed for the development of tomato production (a crop with higher value added) for urban markets, particularly the capital. Large, long-standing patronal farms occupy a whole bank in the valley and thus monopolize most of the water resources with the help of powerful pumping systems and irrigation center pivots. By generating lower values added per surface area unit (wheat, corn) compared to family forms of irrigated agriculture, it is thanks to their very large surface areas that they provide high incomes to their owners (see Chapter 2, this volume).

The lower valley of the Rufiji River has also long been the subject of ambitious planning projects, revived today by the recent arrival of foreign investors. These projects are now presented by their promoters as a privileged path in the context of adaptation to climate change. By greatly reducing the risk of climate change through extensive artificialization of the environment, they would be able to substantially increase agricultural production while securing it in the long term. However, we have seen that the agricultural income of families in the lower Rufiji Valley depends strongly on their ability to combine different cropping systems that take advantage of the environment's different "facets," the floodplain being the most sought after by farmers because of its greater potential, despite the risks involved (see Chapter 3, this volume). However, it is precisely this space that is currently the subject of major planning projects and risks being taken away from the region's inhabitants, resulting in massive eviction. Such a project may therefore result in an increase in populations' vulnerability and precariousness, either by depriving them of access to one of the areas being exploited or reducing all possibilities of compensation and management of the environment's heterogeneity.

7.5 Conclusion

Farmers have been adapting to increasingly restrictive and uncertain climatic conditions for both rainfed and flooded crops (rainfall, length of the rainy season, extent of flooding) for a long time. In all of the regions studied, farmers have been able to evolve their practices and rethink, sometimes in-depth, the functioning of their agricultural production systems in order to cope with a disruption in the climate or a greater frequency of its hazards. To do this, today, as in the past, and in various geographical contexts, they have relied on a detailed knowledge of the ecosystems they use, and expertise adapted to their pedoclimatic conditions, resources, and economic and social context. Common levers for adaptation and increasing their resilience are emerging, often based on an agroecological approach. These include complementary and differentiated use of the different parts of the ecosystem to which they have access, sometimes at the cost of significant artisanal development (gravity irrigation networks, hillside planning), agrobiodiversity reinforcement (crop associations, diversified cropping pattern, mixed livestock farming), increased role of legumes in rainfed agriculture, strengthening links between crops and livestock, and so on.

While these antirisk techniques do indeed fall within the scope of adaptation to climate change or hazards in the various regions studied, they are also at the heart of approaches aimed at better coping with other types of global changes (demographics, prices, integration into market exchanges, etc.). They are often based on labor intensification that makes it possible to compensate for the downward trend in the area farmed per worker, and diversification of production, including that in favor of sales crops if these do not prove too risky. In addition to climatic changes or hazards, these types of farming have been more directly confronted with competition from global markets since the end of the 1980s than during the post-independence period, while receiving less support from national agricultural policies in terms of access to equipment and inputs. They are also more exposed to the risks of land grabbing and the development of forms of agriculture in their direct vicinity with incomparably higher levels of labor productivity.

In the regions studied, the middle and wealthy classes in rural areas, still mainly focused on agricultural activities, have been able to adapt to this combination of changes, often by increasing their agricultural income through higher value-added crops and processing activities, supplementing it with extra-agricultural income. The poorest families, on the other hand, are hard hit by the unfavorable evolution of all these factors (climate, access to land and equipment, relative prices) and some find themselves in situations of great poverty, forced to hire out their hands for part of the year and with no prospect of sufficiently remunerative employment in the expanding urban centers. This reality invites us to not yield to the easy explanation, linking climate change and poverty directly, which is too often repeated in proposed analyses and predictive models, but rather to consider all these global changes' components in order to understand and attempt to contribute solutions to these impoverished and particularly vulnerable farmers.

8 What Policies Should Be Promoted for Adaptation? Lessons from the Past, the Range of Possibilities

Hubert Cochet*, Olivier Ducourtieux, and Nadège Garambois
Comparative Agriculture Training & Research Unit/UMR Prodig, AgroParisTech, Palaiseau, France

8.1 Introduction

The previous chapters have highlighted certain avenues for resilience and adaptation, in line with the extension or renewal of work on vulnerability that highlights the links between poverty, vulnerability, and exposure to risk. These chapters highlight the processes, mechanisms, and trajectories that explain the high exposure of certain groups of farmers and rural people to hazards, as well as the maintenance or increase of this exposure. They help explain the reasons for this vulnerability at the production system level by revealing how the systems operate and the consequences of this, which are particularly economic. They highlight—or illustrate—the influence of past and current choices in terms of agricultural, environmental, and commercial policies on the aggravation of vulnerability, and sometimes, but more rarely, its reduction. Finally, they bring to light the adjustment methods and past and current transformations of agricultural and livestock practices that are moving toward reduced exposure to hazards and a decreased vulnerability in the context of strong demographic growth.

The question now is knowing what lessons can be drawn from these experiences to guide and inform future policy choices. We will first look at the main recent public documents issued by national or international institutions that deal with adaptation. This will be an opportunity to ask ourselves whether the policies for adaptation to climate change, as they are presented, will really change the situation and depart from usual practices. We will then make some proposals for reorienting these policies.

8.2 International Organizations, IPCC, and States: A Brief Status Report on Proposals

In the chapter "Adaptation to and Mitigation of Climate Change in Agriculture" of the World Bank's 2008 report, the issue of adaptation is barely mentioned and only one page is devoted to it. It states "The best way to deal with the increased uncertainty caused by climate change is contingency planning" (World Bank, 2008, p. 242), and the main adaptation measure mentioned is "research and development of crops adapted to new weather patterns and techniques to reduce land degradation" (World Bank, 2008).

---

*Corresponding author: hubert.cochet@agroparistech.fr

©2024 CAB International. *Agrarian Systems and Climate Change: Journeys of adaptation in the Global South* (eds H. Cochet, O. Ducourtieux, and N. Garambois)
DOI: 10.1079/9781800628137.0008

Since then, analyses have been enriched and a large body of gray literature has been produced on the adaptation of agriculture to climate change. Although an exhaustive review of this work is not possible here, we felt it was important to present the main approaches and proposals that come from three types of documents produced by international organizations and states, documents that in the 2010s largely inspire decision making and the public policies implemented. These are respectively:

- the IPCC report (IPCC, 2014)
- reports published by the FAO in 2013 on Climate-Smart Agriculture (CSA) and in 2016 on the state of food and agriculture in the face of climate change
- national documents such as National Adaptation Programmes of Action (NAPAs), National Adaptation Plans (NAPs), or Intended Nationally Determined Contributions (INDCs) developed in the framework of COP 21

8.2.1 IPCC contributions

In their fifth report, published in 2014, particularly *Chapter 7: Food Security and Food Production Systems*, IPCC experts rightly emphasize the complexity of the processes involved and their multi-scalar nature. Drawing from a large body of scientific literature, they also emphasize farmers' adaptive capacity, especially in developing countries, and their past experience taking climate change into account:

> There is increasing evidence that farmers in some regions are already adapting to observed climate changes in particular altering cultivation and sowing times, crop cultivars and species, and marketing arrangements. (. . .) There are a large number of potential adaptations for cropping systems and for the food systems of which they are part, many of them enhancements of existing climate risk management and all of which need to be embedded in the wider farming systems and community contexts (Porter *et al.*, 2014, p. 514).

They insist on the need to consider local knowledge in the search for adaptive solutions, especially in regions where farmers have only difficult and uncertain access to scientific information. They write, for example, in Chapter 22 on Africa:

> Recent literature has confirmed the positive role of local and traditional knowledge in building resilience and adaptive capacity and shaping responses to climatic variability and change in Africa (Niang *et al.*, 2014, p. 1232).

On the other hand, the intertwining of adaptation with climate change, and adaptation with other global changes that farmers face, is repeatedly emphasized:

> Much of the literature covers incremental, reactive adaptation, but given actors are constantly adapting to changing social and economic conditions, incremental adaptation to climate change is difficult to distinguish from other actions, and in fact is usually a response to a complex of factors (Porter *et al.*, 2014, p. 519).

Emphasized as well is the necessary relativization of climate change's future role in relation to the other factors at play:

> It is likely that socioeconomic and technological trends, including changes in institutions and policies, will remain a relatively stronger driver of food security over the next few decades than climate change (Porter *et al.*, 2014, p. 513).

Finally, the experts underline the socioeconomic dimension of vulnerability and the need to go beyond purely technological approaches to adaptation in favor of approaches based on building resilience. They write in Chapter 22, which is devoted to Africa and based on a large body of scientific literature:

> In recognition of the socioeconomic dimensions of vulnerability, the previous focus on technological solutions to directly address specific impacts is now evolving toward a broader view that highlights the importance of building resilience, through social, institutional, policy, knowledge, and informational approaches (Niang *et al.*, 2014, p. 1226).

The fifth IPCC report therefore calls on decision makers to adopt an approach that is not limited to "technical" recommendations but integrates socioeconomic and political aspects, while insisting on the need to value "local knowledge" and striving to strengthen the resilience of the most vulnerable groups.

Recent FAO publications, focused on the CSA approach, unfortunately stray from these recommendations.

8.2.2 FAO recommendations on the CSA approach

8.2.2.1 The concept of CSA

The concept of CSA was formalized in the form of a voluminous manual (570 p.) published in 2013 for practitioners and policy makers. The stated aim is to reconcile three objectives: (i) food security through the necessary increase in food production; (ii) adaptation to climate change; and (iii) climate change mitigation through lower-emission practices (FAO, 2013). The new and laudable ambition of this approach is therefore to integrate the issues of increasing agricultural production and food security into public policies dedicated to climate, rather than treating these two issues separately. The executive summary warns:

> CSA is not a single specific agricultural technology or practice that can be universally applied. It is an approach that requires site-specific assessments to identify suitable agricultural production technologies and practices.

The aim is to increase both the systems' efficiency (the efficiency of the resources used) and resilience, highlighting the diversity of ecosystems at the territorial level, as well as the need to seek solutions adapted to this scale.

However, the approach developed by the FAO is far from the cautious and nuanced recommendations made by IPCC experts on adaptation. The bulk of the report is made up of exclusively technical and sectoral recommendations: "water management," "soils and their management for climate-smart agriculture," "climate-smart crop production system," "climate-smart livestock," and so on. And when it comes to paying attention to the most vulnerable groups (see Chapter 16), recommendations are limited to promoting safety nets and social protection.

As a whole, the report is therefore a vast collection of technical recommendations that, in the interest of a broad consensus, embraces all sorts of approaches, however contradictory they may sometimes be conservation agriculture, integrated farming, organic farming, and so on. Although increasing diversity in the field is emphasized in the first chapter as a pillar of resilience to be promoted,[1] the diversification of production systems is ultimately presented as one option among others. Tissier and Grosclaude, in a critical reading of "climate-smart agriculture (CSA)," write in this regard: "climate-smart agriculture could be more assertive in favor of agricultural production methods that seek to make intensive use of ecosystem services" (Tissier and Grosclaude, 2015, p. 296). They continue, "the designers of CSA do not seem to wish to produce a strong critique of specialized production systems" (p. 297). Thus, the FAO immediately falls into the land-sparing camp (as opposed to land sharing).[2] "Their justification of the most intensive modes of production being used at the global level for mitigating climate change . . . is thus most often nothing more than a cover for 'business-as-usual'" (Tissier and Grosclaude, 2015, p. 296).

Moreover, from the outset, free trade is considered an important element of systems' resilience, with no reference to the sometimes-deleterious consequences of trade's increasing liberalization and its impacts on the least well equipped and least productive form of agriculture. Similarly, no reference is made to the crucial question of which agricultural models should be promoted and the equally serious consequences of the development of farms with large land holdings instead of family farming. Tissier and Grosclaude write on this subject:

> The absence of any reference to family farming in particular—which the overly numerous references to rural communities does not compensate for—will in any case have to be addressed (2015, p. 297).

8.2.2.2 The 2016 FAO report: the state of food and agriculture: climate change, agriculture and food security

In its 2016 report, the Food and Agriculture Organization of the United Nations (FAO) drives home the point of "CSA." The report's main idea is that food security and resilience to climate change can be significantly improved by introducing sustainable agricultural practices, CSA, and techniques and that barriers to their adoption by farmers need to be identified and addressed (FAO, 2016). This is followed by a list of practices and techniques deemed "smart" that should be introduced into agricultural systems. These techniques are classified into broad categories: "sustainable intensification," "agroecology," "efficient water management," "carbon and nitrogen management," and so on. The FAO report is not without contradictions: while agroecology and sustainable intensification are mentioned as solutions, and any form of input subsidy is a priori banned, it nevertheless advocates, for example, for the promotion of a corn technical package (improved seeds and fertilizers) to offset the possible impact of climate change on yields. Four "strategies" for building livelihood resilience are also presented: diversification, support to risk management, reducing gender inequalities, and migration (FAO, 2016, pp. 54–59).

Essentially addressed to developing countries, "climate-smart" agriculture has reportedly generated a definite craze in some of these countries, particularly in Africa and Asia, and led to the formation of a broad alliance to promote it: Global Alliance for Climate-Smart Agriculture (GACSA), launched on 23/09/2014 during the World Climate Summit (Tissier and Grosclaude, 2015).

As far as the African continent is concerned, the initiative for the Adaptation of African Agriculture (AAA) to climate change, launched in the run-up to COP 22 in July 2016, provides an example of this "craze." Topics invoked by pell-mell include integrated fertility management, agroforestry in its "Great Green Wall" component, conservation agriculture, mulch, crop rotations and associations, intercropping, composting, fallowing, the 4% Initiative, smart agriculture and agroecological transition, and so on (AAA, 2016). In seeking to bring together "the ecosystem of existing initiatives," AAA unveils the actors who are rushing to promote "CSA practices": the Soil Fertility Initiative (WB/FAO), Green Revolution in Africa (AGRA), Bill & Melinda Gates Foundation, the Africa Fertilizer and Agro-Business Partnership (fertilizer industry), and so on.

8.2.3   State recommendations

At the country level, numerous reports addressing the issue of adaptation have been produced under the United Nations Framework Convention on Climate Change (UNFCCC).

In the case of the least developed countries, the first was the National Adaptation Programmes of Action (NAPAs) drafted in 2006–2007 and developed to respond to the urgent needs of these countries, at a time when many of them were suddenly affected, in an unprecedented or aggravated way, by increased vulnerability to floods or drought (Least Developed Countries Expert Group, 2012). Then, starting in 2010, National Adaptation Plans (NAPs) were launched, this time designed with a longer-term perspective to take "a more considered approach, working toward transformational change in their capacity to address adaptation . . . reducing vulnerability to the adverse effects of climate change . . ." (Least Developed Countries Expert Group, 2012, p. 15).[3] Burkina Faso approved its NAP in 2015, and Senegal in 2016; many are still being developed (with financial support from UNDP).

A lack of funding has often limited the implementation of NAPA programs, and therefore their real impact. As far as NAPAs are concerned, it is still too early to measure the influence they will have on donors and the degree to which they will actually be implemented but reading them reveals the stated priorities in terms of adaptation policies. The example of Burkina Faso, one of the first countries to have drafted its NAP, illustrates the proposed adaptation measures (Burkina Faso Ministry of the Environment and Fishery Resources, 2015). With regard to agriculture, these include:

- cultivation of early or drought-resistant varieties
- implementation of water and soil conservation techniques (stone barriers, small dykes, small filter dykes, terraces, half-moons, agroforestry, dune fixation, etc.)
- promotion of sustainable land management (SLM)
- improving access to climate information
- implementation of agricultural insurance

  With regard to livestock, the proposals concern:

- fight against bushfires to avoid the destruction of dry-season forage reserves
- implementation of good zootechnical and pastoral practices (pastoral hydraulics, management of pastoral resources, mowing and fodder conservation, forage crops, silage, livestock mobility and transhumance, etc.)
- through capacity building, actors taking climate variability into account in development project and program design

- preservation of cattle breeding severely threatened by climate variability
- farmers' adoption of animal production techniques adapted to the hot climate

As for the actions to be undertaken, a careful reading of these documents reveals that it is most often a matter of recycling projects already under way or already programmed, with various traditional donors (bilateral, multilateral, and NGOs).

A third type of document often reveals the stated priorities for adaptation. These are states' "voluntary contributions" drafted in connection with COP 21 (Intended Nationally Determined Contribution, INDC). Very general considerations are stated, such as: "Promote conservation/smart agriculture activities leading to adaptation benefits and enhancing climate resilience" (Zambia, 2015), but this type of document does not provide any details on the measures advocated.

Thus, whether they are NAPAs, NAPs, or INDCs, these documents drafted by states have in common that they give priority to the formulation of technical recommendations that are very general and rarely regionalized. The adaptation methods implemented by farmers themselves are most often ignored. The role of socioeconomic differentiation in vulnerability is not mentioned, nor is the attention that should be given to the least-endowed farmers.

8.3   Policy Choices that too Often Revert to "Business as Usual"

8.3.1   Technical choices that are often an extension of the Green Revolution

An analysis of the various agricultural policy documents (on adaptation) drafted by states and the recommendations made by international organizations (FAO in particular, GACSA, AAA Initiative) reveals a shared vision of adaptation. Although terms such as "agroecology," "sustainable intensification," "conservation management," and "agroforestry" are used extensively, it is clear that, far from being a real break from previous policies, these documents simply update the recipes and technical packages directly derived from those promoted during the Green Revolution:

improved seeds, fertilizers and pesticides, it being understood that these recommendations are not limited to irrigable regions, as was often the case in Mexico, India, and elsewhere during the Green Revolution. Although the development of shorter cycle or more heat- or drought-resistant varieties is often a necessity, plant breeding, and genetic research are too often *de facto* correlated with the Green Revolution paradigm, as was the search for "high-yielding" varieties at that time.

The case of corn promotion in Zambia and Tanzania (see Chapter 2, this volume) provides an illuminating example of being locked into past technological choices. At a conference in a major hotel in Lusaka in April 2014, experts from USAID/IAPRI and the University of Michigan made their recommendations in these terms: although corn's sensitivity to drought is emphasized and the need for crop and variety diversification is affirmed,[4] "proper timing of fertilizer application and planting high-yielding varieties will remain the largest determinant of yield for most of Zambia" (Olson *et al.*, 2014).

In a context where rapid population growth in many countries of the South, particularly sub-Saharan Africa, makes a significant increase in agricultural production indispensable, decision makers invariably turn to technical models that are likely to increase yields as quickly as possible, particularly those of cereals, in the hope of closing the yield gap that is said to separate yields obtained by farmers from those that would be possible through the straightforward application of a more efficient technical package. Although yield is not the only performance criterion to be taken into account, especially when the technical packages promoted only increase risks to farmers (see Chapter 7, this volume), this attraction for simple, standard models that can be applied under any conditions seems irresistible.

This attraction is all the more important because many studies based on quantitative data processing and modeling methods encourage decision makers to move in this direction. Mathematical modeling imposes simplification: one works with a single crop, grown as a sole crop, often under "average" conditions and applying a "standard" package. The more general the model claims to be, the more it must be based on increasingly simple hypotheses, even if it means erasing all forms of heterogeneity, particularly agrogeographical or socioeconomic.

The work of the IPCC provides a valuable database for this type of work. Combining the scenarios of changes in greenhouse gas concentrations for the 21st century (RCP: RCP 2.6, RCP 4.5, etc.)[5] and various socioeconomic scenarios or "shared socioeconomic pathways" (SSP)[6] opens the way to developing scenarios of climate change's effect on yields, cultivated areas, production, prices, and world markets by 2050 (IFPRI's IMPACT model).

One such model, for Tanzania, was discussed in Chapter 2 (this volume, Ahmed *et al.*, 2011). Based on the various climate models available and on the general equilibrium economic model, the authors attempt to demonstrate (i) that it is climate change, through the decline in cereal yields, that explains poverty; and (ii) that only policies aimed at increasing corn yields—in particular through the technical packages mentioned above—in a context of free competition, are likely to allow for a virtuous "adaptation" to climate change . . . (see Box 8.1).[7]

In the FAO's 2016 report, the CSA version, which has already been cited and is largely devoted to adaptation, much space is given to this type of demonstration. It states for example:

> Simulations using IFPRI's IMPACT model show that the adoption of heat-tolerant crop varieties produces highest projected global yield increase for corn in 205. Varieties that use nitrogen more efficiently produce the highest global yield

increase for rice, while zero tillage is the best option for wheat (FAO, 2016, p. 54).

Or:

> [From Rosegrant *et al.* (2014), and from simulations based on IFPRI's IMPACT model] The number of people at risk of undernourishment in developing countries would be reduced in 2050 by 12 percent (or almost 124 million people) if nitrogen-efficient crop varieties were widely used, by 9 percent (or 91 million people) if zero tillage were more widely adopted, and by 8 percent (or 80 million people) if heat-tolerant crop varieties or precision agriculture were adopted (FAO, 2016, p. 54).

This type of work, using the IPCC's work in a somewhat abusive way is, unfortunately, legion, although IPCC experts themselves warn readers about the precautions to be taken in this matter. . . .[8] This work is partly responsible for the non-contextualized prescriptions that result in standard technical packages with devastating effects. In all cases, farmers are expected to "adopt" the "adaptation techniques" proposed to them. Despite proclamations such as, "there is no standard list of climate-smart agricultural practices that could be applied in all cases" (FAO, 2016, p. 64), the local specificities of each agrosystem are not taken into account, let alone farmers' and herders' unequal access to resources, means of production, and the market. Local expertise and farmers' past experience are not mentioned.

Box 8.1. From mathematical models to promoting adaptation policies. The example of Tanzania. From Ahmed *et al.* (2011).

The stated objective is to examine the link between climate change, changes in corn yield, and poverty, using a quantitative and model-based approach, based on both past and future trends, over a time step of 1971–2031, that is, 60 years.

The model highlights:

– Monthly temperature and precipitation averages for the corn growing season, i.e. January to June, based on data from Phase 3 of the Coupled Model Intercomparison Project (CMIP3) archive of General Circulation Model (GCM) experiments (Meehl *et al.*, 2005) (and considering the SRES A2 emissions scenario; p. 48).

– A link—via multiple linear regression models—between climate change and corn yield based on Tanzanian production statistics for the country's 17 regions, data available for the 1992–2005 period: "The analysis finds that when considering yields as functions of climate, the temperature coefficients are negative, while the coefficients for precipitation are positive (p. 49). Coefficients on both climate variables are highly significant in all models. [. . .] An increase in average growing season precipitation by 1 mm/month is enough to increase corn and rice yields by 0.005 tonnes per hectare."

– The authors do not hesitate to write: "The estimated statistical model, by being based on ex-post data, has the added advantage of endogenizing some adaptive farmer behavior [. . .]. The historical yields thus reflect some adaptability, as do the estimated parameters in the statistical model" (p. 49).

Continued

Box 8.1. Continued.

- To establish the link between changes in corn yields and price changes, the authors use a general equilibrium model with the following assumptions: (i) regarding free and perfect competition: "Since a review of the literature does not offer strong evidence on the nature of competition in the Tanzanian food sector, we opt for the empirically robust assumptions of constant returns to scale and perfect competition here as well. Clearly, this assumption could be altered as more evidence becomes available on the nature of market structures in the food sector" (p. 50); and (ii) about the factor market: "We assume a constant aggregate level of land, labor, and capital employment reflecting the belief that the aggregate supply of factors is unaffected by climate change" (p. 50).
- Finally, to deduce the possible consequences for poverty: "We resolve this complication [...] allowing us to attribute poverty changes solely to climate-based agricultural productivity changes, and not any other event that may cause vulnerability to change between climates in two different periods. Since we are interested in the poverty impacts of interannual variability, we adopt a short-run factor market closure in which land, capital, and natural resources are immobile across sectors. Thus, we assume that a farmer has already made all production decisions under best available adaptive behavior in that timeframe" (p. 51).

The conclusion is clear: "We apply this framework to Tanzania's climate in the 20th century and 21st century, and find that changes in climate volatility are likely to render Tanzanians increasingly vulnerable to poverty episodes through its impacts on staple grains production in agriculture" (p. 53). The authors then justify policies based on increasing yields (corn technical package): "These initiatives to increase agricultural productivity will have to account for changes in the climate more generally, such as through updates to crop calendars and improved crop varieties to account for changing rainfall patterns" (p. 54) and free trade: "In the short run, when resources may not be easily reallocated across economies, open trade regimes have the potential to reduce domestic price volatility" (p. 54). This statement is made after saying that the model does not take global yield increases and their influence on prices, and so on, into account: "In addition, food prices in the United States are not always the same as in Europe. In addition, food prices in Tanzania will be affected to a large degree by changes in crop productivity throughout the world, as these will influence local prices. The current analysis implicitly assumed negligible impacts in other regions, as a way of focusing on the question of how much poverty volatility could be driven by changes in local production" (p. 54).

8.3.2 Priority for large-scale hydraulics, a choice that is always justified?

It is not possible within the framework of this study to draw definitive conclusions on the question of which models should be promoted for water management and irrigation development. However, as this question is raised in most of the regions studied in this research program, some comparison elements can be highlighted, particularly from the perspective of how planning contributes to adaptation to climate change.

It would be difficult to imagine Mekong Delta agriculture without the large planned hydraulic structures that condition usage today. Even if climatic hazards weigh ever more heavily on these structures (risk of exceptional flooding, delta subsidence, and rising sea levels), requiring reinforcement and continuous improvement, there is no doubt that they have been a decisive factor in greatly increasing and securing production as well as raising producers' income.

However, such results would not have been achieved if delta producers did not have shared and secure access to land (agrarian reforms of the 1960s–1970s), means of production, and market under conditions that were themselves secure, while leaving a great deal of autonomy to families in running their farms (see Chapter 4, this volume).

In the Senegal River Delta, several phases of planning (public and then private) succeeded one another, allowing for a continuous increase in flood sheltered and irrigable areas and an improvement in water table control within each hydraulic mesh. There is also no doubt that this fairly large-scale planning, in addition to the spectacular increase in rice production for urban markets, made it possible to overcome the considerable difficulties associated with the climatic deterioration of the 1970s and 1980s and create many jobs, given the possibilities offered by moribund rainfed agriculture and flood recession farming. However, recent extensions of the irrigated area have been entrusted to external investors and the increase in irrigated areas is now generating land conflicts, evictions, and widening inequalities. Under these conditions, will the development of irrigation still be able to play a role in reducing rural people's vulnerability and increasing their capacity to adapt to the changes they face (see Chapter 5, this volume)?

The Lower Rufiji and Kilombero flood valleys in southern Tanzania have been the subject of ambitious planning projects since the early days of colonization, but unlike the Senegal River Valley, these projects did not move beyond the project stage. The recent arrival of new foreign investors and legislative changes facilitating their reception are currently re-launching these projects, now presented by their promoters as a privileged path toward adaptation to climate change. By greatly reducing the climate change hazard through the extensive artificialization of the environment, a substantial increase in agricultural production and long-term security is expected. However, the problem is posed here under conditions that are very different from those that characterized the Senegal River Valley at the time of the first planning. Much more densely populated today, these valleys are now developed by complex, labor-intensive production systems that, in the absence of hydraulic planning, are largely based on skillful management of the environment's heterogeneity (see Chapter 3, this volume). Allocating large parts of these valleys to investors in anticipation of their planning today presents considerable risks for eviction and increased vulnerability, even though their capacity to create a higher value added per unit area than "traditional" production systems remain to be demonstrated over time.

Moreover, in terms of large-scale irrigation, it is not the planned schemes' scope that poses a problem—as demonstrated by the example of the Mekong Delta—provided that effective management and maintenance methods are in place. Often technological choices and the technical packages that accompany them strongly limit potential "beneficiary" farmers' interest in the planning. This was the case in the Senegal River Delta, for example, when the SAED imposed a particularly rigid management sequence on the producers who benefited from public planning. This is still the case today, for example, in the program set up by the KPL company in the

Kilombero Valley in Tanzania, where the management sequence that was popularized—and supposed to increase resilience in the face of climate change—in fact, translates into an increased risk to farmers (see Chapter 3, this volume).

Given the attention—and support—given by public authorities to these large public or private irrigation schemes, it is clear that the development of small-scale irrigation set up at farmers' initiative has gone almost unnoticed. Yet its role is central in the fight against vulnerability and building resilience. In the three situations that we have had the opportunity to study closely, in the highlands of Southern Africa, Zambia, and Tanzania (see Chapter 2, this volume), small-scale irrigation has always proved to be a major lever for increasing farmers' income and production systems' resilience. First, it makes it possible to secure the beginning or end of a "rainfed" crop cycle when rainfall is insufficient or uncertain. It then allows the establishment of off-season crops, which are themselves protected from any moisture stress if water resources are sufficient. In this way, access to small-scale irrigation also makes it possible to broaden the range of possibilities, particularly in terms of higher value-added crops. It also makes it possible to diversify the forage resources available (crop residues in particular) when forage levels are low on rainfed land and thus strengthen livestock production. It is also the preferred way to keep the family labor force busy throughout the year, progressively fill the work calendar, and thus significantly increase labor productivity and income. On the steep hillsides of the Uluguru Mountains (see Chapter 6, this volume) small-scale irrigation, together with the joint use of different ecological stories, has been the backbone of the agrarian system's progressive intensification and its capacity to support very high population densities. Although access to this key resource—irrigation water—is of course very unequal, the fact that these gravity-fed irrigation systems have been built and gradually expanded by farmers themselves, within small-scale collectives, has allowed them to be truly appropriated by their users. The result is relatively efficient resource management. While its use may not be optimal, it produces little conflict and is rather efficient. In the situations studied in this book, very few of these irrigation systems had been supported in any way by public authorities or "projects," with the exception, however, of the market gardening perimeters in the Senegal River Valley.

8.3.3 Despite what is proclaimed, local knowledge and farmers' experience are not sufficiently taken into account

In Chapter 7 (this volume) of the latest IPCC report (*Food Security and Food Production Systems*), the editors rightly emphasize the multiple changes made by farmers to adapt to climate change and the impact of these ongoing adjustments on reducing vulnerability. Based on a review of the scientific literature, they give multiple examples of these adaptive practices: changes in sowing dates, changes in varieties, development of new crops, adjustments in work organization, and so on (Porter *et al.*, 2014, pp. 514–515). These examples are widely illustrated in the previous chapters of this report. They refer to all these practices as autonomous adaptations (or incremental adaptations).[9] They are opposed to "planned adaptations," which refer to larger-scale adaptations, generally by actors outside the rural world (authorities, the public, projects, agricultural entrepreneurs, etc.) who are said to be likely to provoke a real system change toward greater adaptation.

It is the latter type of adaptation that is favored by policy makers and donors, as adaptations implemented by farmers themselves through successive adjustments (incremental adaptations) are judged—when their existence is known—to be largely insufficient, even inoperative. An example of this kind of judgment is given by Mulenga Bwalya (2010, p. 17), concerning Zambian farmers:

> Generally, the results of this particular study indicate that local communities have traditional knowledge and skills to adapt to climate change and climate variability, but these are fairly transitory and less robust to be regarded as adequate strategies to enable them to adapt and mitigate against adverse climate change conditions.

In the national adaptation programs that we were given to consult (NAPA, NAP), little or no reference is made to these "autonomous" adaptation practices; farmers are *de facto* presented as a passive group. This is also the spirit of the FAO report, where adaptation is limited to farmers' adoption of "climate-smart" practices. It is the "barriers to adoption" of these practices that need to be addressed by dedicated policies. Adaptation is conceived/measured as the adoption rate of exogenous innovations: adoption of new crops or varieties, new inputs, and new cultivation techniques. Innovation comes from elsewhere, and so does adaptation, as the adoption patterns set up in the wake of these recommendations replicate the old top-down extension methods that were so ineffective in the past. Since the practices described as "intelligent" are mainly those that come from elsewhere and that we try to popularize in rural areas, what image of "endogenous" peasant practices do we present?

8.3.4 The socioeconomic conditions of the farming profession are not sufficiently considered

The documents mentioned above, which mainly present the range of techniques to be implemented to adapt to climate change, do not fail to point out, in a second step, the obstacles to their "adoption" by rural people. Thus, "land tenure," "economic capacity," "access to technologies," "market access conditions" or "governance" are identified as "barriers to the adoption of adaptation practices" (for example, in Rhodes *et al.*, 2014, p. 25). The FAO, in its 2016 report, proposes "identifying barriers to the adoption of climate-smart smallholder systems" (p. 62). In the report's summary for policy makers, the conclusions states:

> It is crucial, therefore, to strengthen the enabling environment for climate-smart agricultural investments, mainstream climate change considerations in domestic budget allocations and implementation, and unlock private capital for climate-smart agricultural development (FAO, 2016, p. 52).

This "barriers to adoption" approach, which reduces agricultural development to a process of adopting exogenous technologies, is by no means new. As in the past, it is silent when it comes to the modalities of unequal access to land, water, and natural resources, to means of production and markets, and thus on the freedom given to farmers and breeders to change, or not, their practices.

8.3.4.1 The modalities of access to resources (land, irrigation water, pastoral resources) and means of production are not taken into account

The few regions studied in detail in this book provide examples of how farmers are placed in very unequal conditions, depending on their access to resources and means of production, in order to best manage the hazards they face, reduce these hazards' impact, and confront the future challenges of climate change. Sustainable access to sufficient usable farm area, varied ecosystems, seeds, a minimum of inputs, livestock and draft power, and sometimes small irrigation equipment (watering can, hose) can be crucial and make all the difference (see Chapter 6, this volume). However, these aspects are almost always absent from the analyses and proposals formulated for adaptation. The difficulties of access to land experienced by hundreds of millions of poor families are never mentioned as an "obstacle" to adaptation, any more than those related to access to water, when this resource can be mobilized to carry out an off-season cycle. Only "land tenure insecurity" is sometimes mentioned, but this insecurity is limited to the legal aspects of local land tenure systems, while land reforms and titling procedures are invoked to reduce it.

When difficulties in accessing means of production are highlighted as "obstacles to adoption," access to means of production is understood as "access to technologies," which are understood as "modern": motorization, fertilizers, seeds (including GMOs), and pesticides. Institutional barriers and "market failures" then being considered responsible, the "private sector" is called upon to facilitate, in the framework of contract farming, access to technologies for all, as shown, for example, by the FAO (2016) report already cited. It proposes:

1. fostering public–private partnerships to leverage resources
2. designing and piloting innovative investment vehicles that can help attract additional capital by diversifying and managing the risk-return profile of different investors
3. supporting the development and bundling of a wider range of financial instruments, insurance products, warehouse receipts, and value chain finance (p. 119)

To us, this approach seems reductive, to say the least. On the other hand, raising the problem of differentiated access to "means of production," as we propose in this book, immediately raises the question of development inequalities and the methods to implement to deal with them.

8.3.4.2 The real conditions of market access, as well as the consequences of agricultural and food market deregulation, are not addressed

Although proponents of "climate-smart" agriculture observe that a country's overdependence on the international food market can lead to major food crises that hit the poorest populations the hardest (Tissier and Grosclaude, 2015, p. 293), they consider trade to be an essential factor in food system resilience. Indeed, there is no doubt that smooth food trade can contribute to enhancing food security at the regional level, as Mathilde Douillet has demonstrated in relation to corn in Southern and Eastern Africa (Douillet, 2013). However, the policies put forward for adaptation hardly step back from the doxa of free trade, even though numerous scientific studies have shown the considerable impact of setting up competition, without precaution or protection, between agricultures unequally endowed with means of production and with considerable differences in productivity. Ignoring this work, the FAO's profession of faith is unambiguous:

> Trade restrictions, such as tariff and non-tariff barriers, which limit the response of global agricultural production to changes in demand and supply under climate change, should be minimized . . . Climate change is likely to exacerbate existing imbalances between the developed and developing world. [This] underscores the need to help developing countries deal with food and energy price increases, as well as volatility in food supplies (FAO, 2016, p. 102).

The North will feed the South—the die is cast.

8.3.5 Agriculture conceived from the outset as a factor in the degradation of ecosystems

Adaptation policy documents generally make the fight against land degradation—erosion, deforestation, desertification, salinization, and so on—a priority, and "degraded land restoration" projects a major focus of action. As these projects often aim to re-establish plant cover, they are also expected to contribute to significant CO_2 capture and thus mitigation objectives. While significant funds were announced at COP 21 to finance adaptation in developing countries, many projects that could be eligible for this funding claim that their objective is degraded land restoration. However, this objective is not new. In the countries of the South, the colonial era offers many examples of policies that fight against erosion, bushfires, slash-and-burn agriculture, overgrazing, and so on. A review of these experiences is particularly useful at a time when such projects are about to be revived in many places—in the name of adaptation.

In this respect, the example of the early 1950s anti-erosion policy implemented by the British government in the Uluguru Mountains (*Uluguru Land Usage Scheme*), analyzed in Chapter 6, this volume, seemed emblematic. Erosion had long been presented as alarming, although no evidence was shown, and no diagnosis of farming practices was undertaken at the time. The construction of radical terraces was the only protection deemed effective against runoff and landslides. The failure was resounding. The main technique proposed—the radical terrace—was inadequate: difficult to build, with a fragile bank and exposed to erosion, exposing the deeper and less fertile soil horizons, while considerably increasing landslide risk . . . The colonial authorities' obvious failure did not prevent the hillsides of the Uluguru Mountains from being meticulously sculpted with hoes and laid out in ridges, steps, and terraces of various designs by farmers themselves. The agriculture of the Uluguru Mountains offers a spectacular example of farmers' efforts to limit erosion and land degradation and preserve the future, despite the meager means at their disposal.

Although it was carried out for other reasons, the aggressive land reorganization planned and executed in South Africa was also justified in the name of the fight against erosion and overgrazing . . . It offers a dramatic example of an irreversible process of accentuating the phenomena that were supposed to be fought. As soon as the African populations were concentrated in the homelands, on 13% of the country's territory, South African authorities began to plan "agricultural development" reserved for Black people through Betterment Planning programs, in which croplands, pastures, urban housing estates, and woodlands were to be grouped together and obey a standard location pattern (Cochet, 2013). In this context, the prior diagnosis of "native agriculture" by White experts, which emphasized this agriculture's weaknesses (low yields, erosion, overgrazing, etc.), placed the blame on the Bantu populations themselves, even though these weaknesses were the exclusive result of the vast land-grabbing process carried out earlier. This is how the issue of overgrazing became extremely prominent in South African literature. The theme was then used repeatedly to explain the homelands' erosion and degradation and justify the drastically limited number of livestock in Black families' hands. This interpretation overlooked the fact that the concentration of residual livestock in a particularly small area, and the excessive increase in overstocking that it entailed, resulted precisely from the populations' eviction and confinement in reserves. A "tragedy of the commons" was created from scratch, and it came at the right time to justify the energetic fight against the "demon of overgrazing," an additional operation of forced decapitalization. This interpretation also overlooked the fact that for these populations, live cattle represented the best possible way to accumulate capital, which was moreover hindered in a thousand ways. Finally, it ignored the fact that, without any access to synthetic fertilizers, fertility transfers from grazed areas to cultivated fields via livestock represented the only possible mechanism for reproducing the fertility of cultivated areas.

One would like to believe that these policies belong to a bygone era. Unfortunately, this is not the case. Although freed from the deadly ideologies of the Apartheid era, the consequences of land degradation projects and policies may still result in people's eviction, as illustrated by the Ethiopian government's extensive program to set land deemed degraded aside for restoration (Box 8.2).

While there is no doubt that a considerable amount of area is affected by these degradation phenomena, the diagnosis of these processes' causes is too often hasty and generally results in blacklisting "traditional" agriculture and its more or less explicit condemnation, even though a more detailed study of peasant practices often

---

Box 8.2. The Ethiopian government's program: exclosure of land deemed degraded in anticipation of its restoration. According to H. Cochet, 2009.

This land exclosure proposes the exclusion or limitation of human activities and access to livestock on the land concerned. According to the project's creators, it should allow for the reconstitution of plant cover, including woodland regrowth. Through its contribution to increased carbon storage, this program aims to contribute on the one hand to the mitigation of climate change in the perspective of commitments made by the Ethiopian government (INDC), and on the other hand to adaptation. It must be noted that this program is based on a very classic diagnosis, very old and not re-questioned. Thus, it would be human activities, and in particular grazing and the collection of firewood, which, in a context of strong demographic growth, would be solely responsible for land degradation. It is therefore "logical," in order to "restore the environment" and the environmental services attributed to it, to limit human activities in the areas in question. From the "tragedy of the commons" to "land exclosure" decreed from on high, it has been a well-established scenario . . . for centuries.

   This diagnosis is, however, simplistic and cannot by itself be sufficient justification for the degraded land exclosure project advocated by Ethiopian public authorities. The deforestation of a large part of the country's northern highlands dates back several centuries and can in no way be attributed to recent practices of village communities or any contemporary demographic increase. In the southern half of Ethiopia, the phenomenon is less pronounced, and it is now known that certain regions have even experienced a process of afforestation, while others have seen the peasant communities' recent establishment of true coffee agroforests in place of the "deforested" areas that were previously predominant. While it is undeniable that many Ethiopian areas have indeed undergone a disturbing process of biomass decline in recent decades, the phenomenon is complex and multifactorial and calls for an in-depth, case-by-case diagnostic approach.

   The land exclosure policy planned by the Ethiopian government thus appears to be a cure that is worse than the disease. Indeed, most of the areas in question are used by villagers to graze cattle, goats, and sheep for most of the year, with crop residues and (sometimes) purchased food supplementing animals' diets, especially during the dry season. It should be remembered here that the raising of domestic herbivores is a primordial component of the vast majority of Ethiopian agricultural production units and that this livestock activity provides both draft power (bovine draft power is widespread throughout the country) and animal waste. The waste is *indispensable* for the fertility reproduction of cultivated land, when not used, as is the case in the northern half of the country, as fuel when firewood is unavailable. The unilateral (and "planned" by the public authorities) land exclosure of these areas in order to prohibit access to domestic animals therefore presents a *major risk* of irreversibly weakening all the riparian agricultural production units.

---

leads to more nuanced and even diametrically opposed conclusions (Cochet, 2001; Ducourtieux, 2009). Thus, farmers, and in general the poorest among them, are often accused by those in power of being responsible for this degradation. They can thus be the main victims of eviction policies put in place to "stop" this degradation or "restore" the areas concerned. At a time when a new generation of projects to combat land degradation could be created and funded in the name of adaptation, it seems particularly important not to repeat the mistakes of the past.

8.3.6 Conclusion

Unfortunately, the main lines of action for adapting to climate change follow the paths and dead ends of the past, as the lessons of so many repeated failures do not seem to have been learned.

   By focusing all too often on a small number of purely technical recommendations, often all purpose—despite the declared intent of "climate-smart agriculture"—the "adaptation" methods proposed to farmers, when they are not imposed, frequently result in increased risk. Even though the aim is to reduce farmers' vulnerability by limiting the risks inherent in climatic deterioration and increased hazards, it is often the innovation itself that leads to increased vulnerability. The case of unilateral, standard "technical packages" intended for the development of a single crop or variety (corn, rice, etc.), which are supposed to allow for a significant increase in yields but are ill-adapted to both local ecosystems' diversity and farmers' socioeconomic diversity, has been discussed at length in the previous chapters.

As in the past, the fight against "land degradation" still equates too often to authoritarianism and exclosure. As agricultural and pastoral practices are considered archaic and responsible for erosion and deforestation, it is necessary to change them as quickly as possible and impose a "rational" scheme of restoration, protection, and exclosure, even though the resulting evictions/displacements of populations lead on the one hand to increased pressure on neighboring ecosystems and, on the other hand, increased vulnerability of the populations that are victims of all this.

The historical processes of weakening agrarian systems are not studied or taken into account, nor are the mechanisms of differentiation that make it possible to explain the real causes of vulnerability for some, or reasons for other groups' greater resilience, and thus foresee the actions to be undertaken to promote real adaptation. Blindness or power imbalance? It is as if the diagnoses of vulnerability deliberately ignore the weight of the social relationships and social groups that have inspired past political choices. By reducing the scope of the diagnosis and the proposed solutions to an enumeration of technical recommendations, the question of adaptation escapes any political questioning. Curiously, while the causes of climate change are now recognized by (almost) everyone as being mainly of anthropogenic origin, the climate "hazard" and risks to farmers are still implicitly presented as "natural." And "climate change" thus becomes, as if by magic, the number one (or even the only) threat to farmers in the most exposed regions.

However, vulnerability is above all a social construct, as we have shown through the diachronic and typological approach developed in each of the regions studied in this research work. This is why the definition of policies to reduce vulnerability, with the prospect of better adapting to climate change, presupposes that clear political choices are made in favor of the most vulnerable groups. However, financing available in the name of adaptation is more likely to lead to mechanisms for capturing opportunities that benefit those in the best position, as Vanderlinden suggests:

> The world in which these opportunities will unfold is not a "flat" world. It is a world in which power relationships already exist and where those who are already well placed will be in the best position to seize, or even—and this is the most worrying—contribute to the definition of these "benefits" (the expected benefits of mitigation policies)[. . . If these same groups are called upon to contribute to the definition of adaptation modalities, will they choose to actively fight against inequalities or will they instead encourage distributed technical solutions via a market that already favors them? (Vanderlinden, 2015, pp. 41–42).

8.4 A Necessary Renewal of Approaches and Putting Policy Proposals to the Test by Facts

8.4.1 Truly relying on farmers' expertise

The extensive scientific literature on agricultural adaptation with limited resources and means of production has had the merit of recognizing certain "traditional" techniques' effectiveness in terms of adaptation. This is the case, for example, for Zaï in the Sahelian zone, *Faidherbia albida* tree stands in the Sahelo-Sudanian zone, or even, more simply, crop-livestock mixing. The belated recognition of these practices' effectiveness is, however, insufficient, insofar as it has not been able to significantly influence the policies promoted in the name of adaptation.

The situations studied in detail in the preceding chapters also provide numerous examples along the same lines: management of the dual rainfall and flooding hazard by combining rice and corn in the low Kilombero and Rufiji floodplain valleys (Tanzania) (see Chapter 3, this volume); selection and association of several crop species and varieties, especially cowpeas, in the northern Senegalese Peanut Basin (see Chapter 1, this volume); associated cropping and small-scale irrigation in the Zambian and southern Tanzanian highlands (see Chapter 2, this volume); cropping associations and successions on the stepped hillsides of the Uluguru Mountains (see Chapter 6, this volume); and combinations of several varieties and rice-growing systems in the Tonle Sap Basin in Cambodia (see Chapter 4, this volume). In addition, it seems essential to draw on the particularly rich experience of farmers who have already experienced phases of marked climate deterioration in the past (Sahel in particular, but also Tanzania at the very time the villagization policy was being implemented). The preceding chapters show how the modalities of adaptation to these old climate changes (changing crops and varieties, modifying cropping calendars, animal husbandry/mobility, etc.) enrich today's search for solutions. These old situations also make it possible to identify which production systems were most resilient to climatic deterioration and which farmers managed to cope with or suffer less from it. Likewise, the situations make it possible to highlight which groups were the most vulnerable, explain the reasons for this, and follow the groups' trajectories.

Relying on farmers' experience, especially the poorest among them, is also particularly rich in lessons. Indeed, because they were barred from all access to means of production—synthetic fertilizers, pesticides, "improved" seeds—they are the ones who have implemented low-input solutions, sometimes the most diversified, and, provided they are supported by appropriate policies, are today best able to develop agriculture that is adjusted to environmental conditions, contributes very little to GHG emissions, and is capable of adapting. Koohafkan and Altieri (2011) also emphasize the adaptive potential of "traditional" agriculture linked to its high level of biodiversity.[10]

Rather than getting bogged down in the search for "climate-smart techniques" and trying to remove obstacles to their massive adoption by farmers, a top-down process doomed to failure, effective adaptation policies should therefore first and foremost seek to identify the endogenous processes that farmers have put in place (or that they would like to develop) in order to reduce risk, so as to build on these processes and try to increase their scope and effectiveness through appropriate measures.

8.4.2 Promoting diversity

While crop diversity at the field level can be a guarantee of risk dispersion and greater resilience, as we have seen above (varietal diversity, associated crops), it is at the farm level that this diversity reveals its greatest potential. Specialization in a single crop or variety is always extremely risky from a climatic point of view, unless the environment is highly artificial (controlled irrigation, leveling, etc.). It is also risky from an economic point of view, except in the increasingly rare situations where farmers are protected from excessive price fluctuations and sure to receive sufficient income from selling their harvest to meet their needs. On the contrary, using a diverse mix of crop species and varieties is the best way to reduce risk and ensure that, whatever the circumstances, a minimum crop will still be obtained. Allowing for the selection of the most suitable environment for each plant and the most suitable plant(s) for each area is often also the least expensive technique to implement, as the increased use of the environment's functionalities often allow one to limit the input doses used. Finally, a varied mix of crops and livestock is also the best way for a farming family to keep the family labor force—whose opportunity cost is often very low due to the lack of real employment alternatives—busy for as long as possible throughout the year (filling in the work calendar) and thus increase its overall productivity.

This is why it is imperative to abandon programs and projects aimed at specializing farmers in one or too few crops or varieties instead of more diversified systems. Similarly, insofar as access to diverse ecosystems considerably facilitates the implementation of diversified, more productive, resilient production systems, it seems imperative to put an end to uncoordinated land exclosure projects—even in the name of adaptation—which would deprive farmers of their fodder resources. It is also important to stop the processes of eviction that result from land grabbing in some places and amputate village territory, thereby restricting farmers' access to certain ecosystems.

8.4.3 Thinking about adaptation at the farm level, not only the field level

The main policy documents devoted to adaptation to climate change most often present, *in fine*, a catalog of "adaptation techniques" to be promoted, most of which are applied at the field level. In addition to the Green Revolution-type technical packages that have been widely discussed and whose inadequacy we have emphasized, recommendations that are closer to agroecological practices (agroforestry, mulch, conservation agriculture, application of compost, etc.) are also proposed and evaluated at the field scale.

However, their true effectiveness, or, on the contrary, the danger they may represent, can only be perceived and analyzed at the scale of the entire farm, at the production system scale.

The example of mulching coffee fields in Burundi proves enlightening. Mulching is unanimously recognized as a virtuous practice, both from a mitigation point of view (recycling of organic matter as a substitute for synthetic fertilizers, etc.) and from an adaptation point of view. It is known to be a keystone of so-called "conservation agriculture." But in Burundi fields, it turned out to be a truly destructive practice for both the country's food-producing agriculture and its export revenue, which mainly comes from coffee. The systematic mulching of coffee trees has in fact monopolized almost all of the available biomass, at a time when its opportunity cost has continued to rise. It has led to an accelerated decline in the fertility of fields that provide biomass, particularly for food crops, and has led to a serious crisis on many farms (see Box 8.3).

The technique of composting, also popularized in the name of adaptation in many countries today, must be analyzed in the same way. Compost is of course beneficial to the field to which it is applied, but what are the results on the whole farm or even agricultural system scale? While the carbon balance is largely positive if the biomass used for composting is burned in the field when there is no compost pit, is the composting facility beneficial to the farm as a whole if the same residues are buried at plowing time or consumed by domestic animals in periods of fodder scarcity?[11] Thus, for example, in looking at farms declared "climate-smart" because they have "adopted" the composting technique, Andrieu and colleagues find that when there are insufficient crop residues, the introduction of compost on farms results in increased transhumance duration among Burkinabé herders (Andrieu *et al.*, 2015).

Based on zero tillage and the maintenance of the previous cycle's crop residues, conservation agriculture, although presented as one of the possible ways to reduce cropping system vulnerability, is confronted with the same design flaws. Closing the analysis scale off to the field level and insufficient consideration of the promoted technique's real effects on the farm level, as well as the diversity of situations in which farmers find themselves, explain the relative failure of many projects in this area (Levard *et al.*, 2014; Dugué *et al.*, 2015).[12]

Finally, all the techniques promoted for water and soil conservation management or the fight against land degradation should also be evaluated, on a case-by-case basis, taking into account the opportunity costs of the resources and work time devoted to them.

Box 8.3. How a technique considered relevant at the field level turns out to be detrimental at the farm level: the example of mulching coffee trees in Burundi. According to Cochet (2001).

Since the first plantations in the 1930s and as the national orchard expanded, mulch has always been intensively popularized and closely supervised by producers. Its agronomic functions are as follows: it slows down the soil's drying during the dry season; as it decomposes progressively, it enriches the soil with humus; it prevents weeds from developing; and it provides effective protection against rain erosion and runoff.

Although this technique's effectiveness is well known by farmers—they obtain better coffee yields when they have mulched their coffee fields regularly, abundantly, and in a timely manner—its "adoption" has always required particularly authoritarian "supervision" by agricultural services.

The main reason for this is that organic matter does not fall from the sky . . . It mostly comes from crop residues (banana leaves and stripes; corn and sorghum residues; bean, sweet potato, or peanut haulm), weeds (including quack grass roots), and sometimes residual pastures, which are then enclosed and mowed. This biomass must be harvested and then transported to the coffee field for spreading, the optimal quantity being 20 to 25 t of dry matter (DM) per hectare per year.

However, many farmers are unable to gather such a large amount of organic matter each year due to a generalized biomass shortage, largely because of the regular decrease in average farm size (estate divisions), increase in the ratio of area needing mulching vs mulch-producing area, decrease in observed food crop yields (residue amounts are also affected), and decrease in or near disappearance of permanent grassy areas inside or outside farms.

Continued

Box 8.3. Continued.

By finally causing a massive organic matter concentration on the coffee field, mulching enriches the coffee plantation with humus and fertilizing elements. This transfer is achieved at the expense of the fertility of other fields on the farm, whose fertility decreases. Instead of being buried during plowing, the cereal and legume crop residues are devoted to coffee. Instead of mulching the banana plantation and returning part of the biomass produced to the soil, banana leaves and stripes are gathered in bundles and transferred to the coffee. If the coffee is mulched by mowing the residual pastures, it is because they have literally been enclosed, forbidden from grazing, and reserved for mulching coffee.

Mulching coffee trees thus results in massive fertility transfer to the detriment of food crop fields, banana plantations, and residual pastures. By accentuating the fertility decrease in these fields, the obligation to mulch weakens the farm economy and leads to a decrease in labor productivity. Thus, the systematic and unilateral popularization of a technique "adapted" to coffee—but not to the entire farm—has led to an agronomic, economic, and social impasse.

8.4.4 Promoting shared access to means of production

The work carried out within the framework of this research program largely illustrates the extent to which access to the means of production has been crucial in historical periods marked by climatic deterioration and how it is still essential today. As we have seen, it is not a question of simple "access to technologies," reduced to its dominant technical conception, that would come with the popularization of this or that technical package designed in experimental research stations. It is a question of all farmers, including the poorest, having access, under decent conditions, to the means of production adapted to their needs and local production conditions.

8.4.4.1 Diversified plant material and inputs adapted to local needs

For each region studied, the preceding chapters provide a fairly precise idea of the main production systems present, as well as the annual productivity and income levels, including on-farm consumption, which can be identified. The orders of magnitude to retain for the most numerous farmer categories are frequently around a few hundred euros per worker. These very low incomes (total agricultural income, including on-farm consumption of home-produced foods and meat) sometimes result in great food insecurity and paltry purchasing power and investment capacity. This state of affairs is as much the cause as the result of limited or non-existent access to the most basic means of production, a handicap that makes these farmers very

vulnerable, particularly in relation to climatic hazards. We know that this extreme precariousness sometimes forces some farmers to sell their entire food crop to pay off their debts and then buy back the necessary basic commodities at high prices on a daily basis. Sometimes, families have such difficulty in obtaining the necessary seeds each year that the areas sown with a particular crop depend on the opportunities encountered for certain seeds, which makes it impossible to establish the crops and crop associations best suited to each location and each season. While sowing is done as best as possible according to the opportunities seized to obtain seeds at the last moment, the crop cycle is shifted, which leads to both further yield losses and a shift in the next cycle in regions where it is possible to carry out two cycles in the same year. The farm is then locked into a cycle of precariousness and crisis from which it is sometimes impossible to escape.

Malawi and, to a lesser extent, Zambia have, against donor advice, reintroduced vigorous input subsidy policies, particularly for synthetic fertilizer. Although these policies contributed to locking farmers into overly rigid technical packages based exclusively on sole cultivation (see Chapter 2, this volume), there is no doubt that they were successful when these inputs were made available in a timely manner. This was also the case in the Senegalese Peanut Basin in the 1960s and 1970s. This is why it is surprising to see the FAO still maintaining an outdated position when it unceremoniously classifies any form of input subsidy as a "barrier to the adoption of improved practices" (executive summary of the *State of Food and Agriculture* Report 2016: *Climate Change, Agriculture and Food Security*).

The statement is more nuanced in the overall report, however:

> In developing countries, trends are toward increasing use of producer price support and input subsidies. Input subsidies are often motivated by the belief that, by reducing input costs, yields will increase and food security will improve . . . Hence, careful assessment and policy design are needed to avoid creating incentives that counteract environmental goals. One way of aligning agricultural development and climate goals would be by making agricultural support measures conditional upon the adoption of agricultural practices that lower emissions and conserve natural resources (FAO, 2016, p. 98).

Facilitating access to diversified plant material (seeds, cuttings, seedlings, scions, etc.) and inputs adapted to farmers' needs and the ecosystems in which they work is crucial. Provided, of course, that we renounce imposed (or "proposed") technical packages, and we strive to ensure that these newly accessible means of production facilitate and accompany the implementation of agroecological practices and do not lead to simplifying rotations, crop associations, and management sequences.

8.4.4.2 Facilitating access to livestock and appropriate tools

The same is true for access to tools and equipment, as well as for capital held in livestock reserves. The previous chapters have shown how draft animal power and a complete set of equipment (seeder, weeder, cart) enabled Senegalese farmers who had access to it to cope with the shortened cropping seasons in the 1970s as well as the urgent need to plant millet and peanuts more quickly and increase weeding in an extremely short period of time (see Chapter 1, this volume). The development of using tractors as a service provision for tilling is a step in the same direction, provided that their use is compatible with farmers' purchasing power and does not increase their debt. In Cambodia, in the Tonle Sap Basin (see Chapter 4, this volume), the acquisition of a power tiller has enabled some families to save precious time at the beginning of the cycle: the speed of motorized plowing, combined with the possibility of dry soil tilling, makes it possible to extend the calendar window for

tilling by avoiding the first rains, while at the same time significantly reducing risk. In the Mekong Delta (Vietnam), the planned and State-subsidized introduction and gradual popularization of multi-purpose power tillers (transport and plowing) during the 1980s–1990s, as well as the introduction of the first tractors, made it possible to switch to three annual rice cycles, as soon as hydraulic planning, land access conditions (agrarian reforms), and availability of inputs made it possible to do so (see Chapter 4, this volume).

In the highlands of Zambia and southern Tanzania, but also in the Uluguru Mountains (Tanzania), we have demonstrated the crucial importance of small-scale irrigation equipment, however modest—rubber hose, watering cans, small treadle pumps, and so on—in the development of off-season crops (see Chapters 2 and 6, this volume).

We can therefore see that facilitating access to agricultural tools and equipment, but also to storage and post-harvest processing equipment, means of transport (carts, which are essential for any form of biomass transport), and of course livestock (since mixed crop-livestock practices have proven their worth in many contexts), should be part of any policy to increase resilience and help adapt to climate change.

8.4.5 Conditions to be met

We have emphasized that vulnerability is above all a social construct, resulting from processes of socioeconomic differentiation that are here for the long term. Any adaptation policy—and therefore first and foremost reducing vulnerability—must therefore be part of a plan to fight against inequality in development and before all else, inequalities in access to resources. We share the idea that adaptation can only be considered in terms of reducing inequality:

> There is a broad consensus that inequality is closely linked to vulnerability/exposure, and this predates the issue of climate change. But the emphasis on inequalities in access to resources—including education as an enabling factor for the invention and implementation of adjustment mechanisms—is the result of more recent theoretical and empirical work. Inequality in access to resources creates an

adaptive capacity deficit for the least advantaged groups, a deficit that can potentially destabilize society as a whole. As a result, adaptation today can only be thought of in terms of reducing inequalities, regardless of the dimension concerned (Vanderlinden, 2015, p. 41).[13]

A situation often advocated (for example, in the 2008 World Bank report) is the exit of all "surplus" rural people from the agricultural sector, thus freeing up the factors of production (notably land) to be allocated to the highest performers. On the contrary, it is essential to promote inclusive development for all, therefore including those who are currently in the most difficult situations and are also the most vulnerable to climate change. Although income sent by migrants sometimes plays a decisive role in the survival of the households that remain in the village, sometimes even making investments that increase farm resilience, exiting the agricultural sector (migration) should not be understood as a climate change adaptation strategy, as one can sometimes read in the same authors' writings, but rather as evidence of extreme vulnerability and difficulties adapting.

8.4.5.1 *Democratizing access to resources (especially land and water)*

The first resource that must be ensured is land. In all the regions studied in detail in the previous chapters, the most vulnerable are always those whose access to land is the most limited and precarious. Thus, there can be no realistic adaptation policy that does not take place in a context of greater land access equity. In many situations, it is indeed access to this resource that represents the greatest uncertainty for the most vulnerable groups, either because they were excluded in previous periods or because they do not have access to it today (particularly young people and women). Thus, although the issue of land tenure goes well beyond the scope of this book, it cannot be ignored or limited to the question of "securing" land through property titling procedures.[14]

It goes without saying that such a policy must take a firm stand in favor of family farming and reflect a clear choice in terms of development models that allow available resources to be allocated to family farming and not to projects whose exclusive effects, including in terms of social eviction, are obvious.

8.4.5.2 *Reinventing new forms of market regulation*

Finally, one of the main hazards facing farmers is the level of agricultural prices (input prices, prices of agricultural commodities) and their very strong fluctuations, which immediately raises the question of trade policies and, with them, that of the free trade agreements in which the countries of the South are involved. It seems difficult to talk about reducing vulnerability without thinking about the steps needed to reduce economic hazards and risks, in conjunction with climatic hazards and risks. From this perspective, the example of Vietnamese policy mentioned in Chapter 4 (this volume) of this report is instructive.

8.5 Conclusion

In an attempt to reduce farmers' vulnerability and increase their resilience, we must rely on farmers themselves; that is, on their in-depth knowledge of the ecosystems that make up both their living environment and way of working. Their capacity for innovation needs no further proof, nor does their ability to discern between innovations that increase vulnerability and those that reduce it. In our view, the priorities therefore seem to be:

- In each region, identifying the dynamics under way, as well as the practices and innovations showing both a risk-limitation strategy and an increased resilience capacity, particularly regarding biomass management and storage.
- Supporting farmers through programs that facilitate access to the means of production they need and by securing the conditions of this access, including by intervening, when necessary, in the relative price system.
- Reducing risk in all areas possible especially when public action itself is at the source of risk, uncertainty, and vulnerability for the most modest producers; for example, as in the land tenure system, or the authoritarian popularization of technical packages that carry risks and are imposed from above.

In the current debate on adaptation policies, it is often said that there are progressive and perceptible adaptations at the farmer practice level on the one hand (incremental changes or incremental adaptation) and more radical transformations requiring a real system change on the other (transformative responses or transformative adaptation). The latter is often favored by decision makers for the sake of the urgency and efficiency expected from policies. Thus, in Chapter 20, entitled *Climate-Resilient Pathways: Adaptation, Mitigation, and Sustainable Development*, the latest IPCC report writes:

> Incremental responses are often referred to as business-as-usual approaches, as they do not challenge or disrupt existing systems (. . .) Transformative responses, in contrast, involve innovations that contribute to systemic changes by challenging some of the assumptions that underlie business-as-usual approaches (Denton *et al.*, 2014, p. 1106).

Our approach to the issue of adaptation places a premium on incremental responses implemented by farmers themselves. However, it would obviously be wrong to deduce, as this passage from the IPCC report suggests, that this approach would favor a "*laissez-faire*" position, let alone "business as usual." On the contrary, the measures that we propose to accompany and reinforce these incremental responses are a major departure from policies promoted in the past and those that are now being promoted in the form of transformative responses.

Indeed, thinking about real climate change adaptation policies in a context of strong demographic growth, scarcity of certain resources, and marked increase in inequalities (particularly concerning access to resources), while taking into account the need to reduce agriculture's contribution to climate change (mitigation), means rethinking agricultural and rural development policies in depth. There can be no question of limiting ourselves to a single "adaptation component" of a climate policy, however ambitious it may be. Reconciling adaptation, mitigation, and food security, as proposed by those who have rallied around the Climate-Smart Agriculture initiative, is in itself a laudable objective, as is the necessary strengthening of various sectoral policies' coherence (C2A, 2017). Meeting this challenge will only be feasible by resolutely tackling the mechanisms that create inequalities or evictions when it comes to access to resources and means of production, and reinventing market regulation tools that allow farmers to make a decent living from their activity and invest. This will only be possible by promoting agriculture that provides employment, is as diversified as possible, and makes only sparing use of synthetic inputs; in short, drawing the curtain on developing agriculture based on a small number of technical packages or "adaptation techniques" that are conceived from on high and unsuited to farmers' very real circumstances.

Notes

[1] "Increasing diversity of production at farm and landscape level is an important way to improve the resilience of agricultural systems" (FAO, 2013, p. 23).

[2] The "land sparing" approach gives priority to major intensification on limited surface area in order to occupy as little space as possible and leave as much area as possible dedicated to "protection." The "land sharing" approach, on the other hand, proposes relying on the ecosystem's functionalities and complementarities to limit the use of synthetic inputs, even if it means considering the farmers' environment as a multifunctional whole without separating productive spaces from spaces dedicated to conservation.

[3] See the C2A report (2017).

[4] "With more temperature and rainfall extremes and variability especially in the South, East and West, reducing vulnerability will require diversification of crops and varieties, and good soil and water management (e.g. agroforestry, aspects of conservation agriculture)" (Olson *et al.*, 2014).

[5] RCP: Representative Concentration Pathway (representative profiles of concentration evolution defined by the IPCC).

[6] The Shared Socioeconomic Pathways (SSPs 1–5) describe different plausible hypotheses for future societal and ecosystem developments over the 21st century. They are used alongside the RCPs to analyze the reaction between climate change and factors such as world population growth, economic development, and technological progress. They are based on scenarios describing possible developments that pose different challenges for adaptation and mitigation (FAO, 2016, p. 35).

[7] Another example of this type of modeling done by IFPRI researchers is given by Jalloh *et al.* (2013) for the case of West Africa. See also the acritical literature review proposed by Rhodes *et al.* (2014) in which numerous studies of this nature are cited.

[8] The IPCC report insists, after explaining the methods and models used, on the caution to be exercised with regard to these projections. It specifies that in the last 50 years, studies linking climate change and yield evolution have not been based on any counterfactual scenario: "Many studies of cropping systems have estimated impacts of observed climate changes on crop yields over the past half century, although they typically do not attempt to compare observed yields to a counterfactual baseline" (Porter *et al.*, 2014, p. 491). He also points out the difficulties that arise when extrapolating experimental results obtained in the USA and China to the intertropical zone: "Unfortunately, the FACE (Free Air Concentration Enrichment) experiments are carried out mostly in the USA and in China, and thus limited to specific environmental conditions, which do not fully reflect tropical or subtropical conditions, where CO_2 and soil nutrient interactions could lead to large differences in photosynthesis rate, water use, and yield" (Porter *et al.*, 2014, p. 495) [. . .] "Also, the number of FACE studies is still quite low, which limits statistical power when evaluating the average yield effects of elevated CO_2 or interactions with temperature and moisture" (Porter *et al.*, 2014, p. 495) and again: "Given these different strengths and weaknesses, and associated dependencies, it is critical that both experimental and modeling lines of evidence, and their uncertainties, are examined carefully when drawing conclusions regarding impacts, vulnerabilities, and risks" (Porter *et al.*, 2014, p. 496).

[9] Defining them as follows: "Autonomous adaptations are incremental changes in the existing system including through the ongoing implementation of extant knowledge and technology in response to the changes in climate experienced. They include coping responses and are reactive in nature" (Porter *et al.*, 2014, p. 513).

[10] In this regard, the IAASTD (International Assessment of Agricultural Science and Technology for Development) report, approved by the intergovernmental plenary assembly in April 2008, rightly insisted on the importance of supporting the most modest farmers, the complex and multifunctional nature of agriculture, and the importance of valuing local expertise (IAASTD, 2009).

[11] To be complete, the balance with/without a compost pit should of course include the opportunity cost of the work required to make and spread the compost: collection of organic matter, watering, turning, transport and spreading, and so on.

[12] Significant programs to promote conservation agriculture have been implemented in Zambia, for example, whose successes and limitations have been discussed by Leménager and Ehrenstein (2016).

[13] Hansen and his colleagues also consider that "improving rural livelihoods now through aggressive pro-poor development may be the most promising avenue for adapting to future climate change" (2007).

[14] On these questions, we refer the reader to the work of the Land Tenure and Development Technical Committee (GRET/AFD) and to the abundant literature devoted to it.

References

AAA (Adaptation of African Agriculture) (2016) *The Initiative for the Adaptation of African Agriculture to Climate Change (AAA): Addressing the Challenges of Climate Change and Food Insecurity.* White paper, 24 pp.

Ahmed, S.A., Noah, S., Diffenbaugh, N.S., Thomas, W., Hertel, T.W. *et al.* (2011) Climate volatility and poverty vulnerability in Tanzania. *Global Environmental Change* 21, 46–55.

Andrieu, N., Pédelahore, P., Howland, F., Descheemaeker, K., Vall, E. *et al.* (2015) Exploitations agricoles climato-intelligentes? Etudes de cas au Burkina Faso et en Colombie. In: Torquebiau, E. (ed.) *Changement climatique et agricultures du Monde.* Quae, Paris, France, pp. 136–146.

Burkina Ministère de l'environnement et des ressources halieutiques (2015) Plan National d'Adaptation aux changements climatiques (PNA) du Burkina Faso. Volume principal 6/2015, Ouagadougou, Burkina Faso, 155 pp.

C2A (2017) *Quelles politiques publiques pour promouvoir l'adaptation des agricultures familiales aux changements climatiques.* Rapport de la C2A, Coordination SUD, Commission Agriculture et Alimentation (C2A), 82 pp.

Cochet, H. (2001) *Crises et révolutions agricoles au Burundi.* Karthala/INA P-G, Paris, France, 468 pp.

Cochet, H. (2013) *La question agraire en Afrique du Sud: échec d'une réforme.* Focales N° 17, AFD, Paris, France, 158 pp.

Denton, F., Wilbanks, T.J., Abeysinghe, A.C., Burton, I., Gao, Q. *et al.* (2014) Climate-resilient pathways: adaptation, mitigation, and sustainable development. In: *Climate Change 2014: Impacts, Adaptation, and Vulnerability. Part A: Global and Sectoral Aspects. Contribution of Working Group II to the Fifth Assessment Report of the Intergovernmental Panel on Climate Change*. Cambridge University Press, Cambridge, UK, pp. 1101–1131.

Douillet, M. (2013) Maïs en Afrique de l'Est et Australe: la sécurité alimentaire régionale liée à l'amélioration du fonctionnement des marchés. Le Déméter, pp. 205–226.

Ducourtieux, O. (2009) *Du riz et des arbres: L'interdiction de l'agriculture d'abattis-brûlis, une constante politique au Laos*. Karthala/IRD, Paris, France, 371 pp.

Dugué, P., Djamen Nana, P., Faure, G. and Le Gal, P.Y. (2015) Dynamiques d'adoption de l'agriculture de conservation dans les exploitations familiales: de la technique aux processus d'innovation. *Cahiers Agriculture* 24, 60–68.

FAO (2013) *Climate-Smart Agriculture Sourcebook*. FAO, Rome, Italy, 570 pp.

FAO (2016) *The State of Food and Agriculture: Climate Change, Agriculture and Food Security*. FAO, Rome, Italy, 191 pp.

Hansen, J.W., Baethgen, W., Osgood, D., Ceccato, P. and Ngugi, R.K. (2007) Innovations in climate risk management: protecting and building rural livelihoods in a variable and changing climate. *Journal of Semi-Arid Tropical Agricultural Research*.

IAASTD (International Assessment of Agricultural Science Knowledge and Technology for Development) (2009) *Agriculture at a Crossroad*. Global Report, McIntyre, B.D. (ed.) (IAASTD, secretariat), Herren, H.R. (Millennium Institute), Wakhungu, J. (African Centre for Technology Studies) and Watson, R.T. (West Anglia University). Island Press, Washington, DC.

IPCC (2014) *Climate Change 2014: Impacts, Adaptation, and Vulnerability. Part B: Regional Aspects. Contribution of Working Group II to the Fifth Assessment Report of the Intergovernmental Panel on Climate Change*. Barros, V.R., Field, C.B., Dokken, D.J., Mastrandrea, M.D., Mach, K.J. *et al.* (eds). Cambridge University Press, Cambridge, UK, 688 pp

Jalloh, A., Nelson, G.C., Thomas, T.S, Zougmoure, R. and Roy-Macauley, H. (eds) (2013) *West African Agriculture and Climate Change*. IFPRI, Washington, DC.

Koohafkan, P. and Altieri, A. (2011) *Globally Important Agricultural Heritage Systems: A Legacy for the Future*. GIAHS/FAO, Rome, Italy, 41 pp.

Least Developed Countries Expert Group (2012) *National Adaptation Plans. Technical Guidelines for the National Adaptation Plan Process*. UNFCCC Secretariat, Bonn, Germany, 148 pp.

Leménager, T. and Ehrenstein, V. (2016) Des principes agroécologiques à leur mise en pratique. Quels effets environnementaux en Zambie et quels enseignements pour les bailleurs de fonds? *Revue Tiers Monde* 65–93.

Levard, L., Vogel, A., Castellanet, C. and Pillot, D. (2014) *Agroécologie: évaluation de 15 ans d'actions d'accompagnement de l'AFD*. Synthèse du rapport final. AFD/GRET, Evaluation de l'AFD n°58, Paris, France, 20 pp.

Meehl, G.A., Covey, C., McAvaney, B., Latif, M.M. and Souffer, R.J. (2005) Overview of the Coupled Model Intercomparison Project. *Bulletin of the American Meteorological Society*, 86, 89–93.

Mulenga Bwalya, S. (2010) *Climate Change in Zambia: Opportunities for Adaptation and Mitigation through Africa Bio-Carbon Initiative*. CIFOR, Lusaka, Zambia.

Niang, I., Ruppel, O.C., Abdrabo, M.A., Essel, A., Lennard, C. *et al.* (2014) Africa. In: *Climate Change 2014: Impacts, Adaptation, and Vulnerability. Part B: Regional Aspects. Contribution of Working Group II to the Fifth Assessment Report of the Intergovernmental Panel on Climate Change*. University Cambridge Press, Cambridge, UK, pp. 1199–1265.

Olson, J., Crawford, E., Wineman, A., Mulenga, B., Chabala, L. and Kuntashula, E. (2014) *Climate Change Impacts on Agriculture and Household Decisions, and Implications for Adaptation*. USAID/IAPRI/ Michigan State University, Presentation at Chrismar Hotel, Lusaka, Zambia.

Porter, J.R., Xie, L., Challinor, A.J., Cochrane, K., Howden, S.M. *et al.* (2014) Food security and food production systems. In: *Climate Change 2014: Impacts, Adaptation, and Vulnerability. Part A: Global and Sectoral Aspects. Contribution of Working Group II to the Fifth Assessment Report of the Intergovernmental Panel on Climate Change*. Cambridge University Press, Cambridge, UK, pp. 485–533.

Rhodes, E.R., Jalloh, A. and Diouf, A. (2014) Revue de la recherche et des politiques en matière d'adaptation au changement climatique dans le secteur de l'agriculture en Afrique de l'Ouest. IDRC/CRDI, Future Agricultures, Document de travail 090, 56 pp.

Rosegrant, M.W., Jawoo K., Cenacchi, N., Ringler, C., Robertson, R., Fisher, M., Cox, C., Garrett, K., Perez, N.D. and Sabbagh, P. (2014) *Food security in a world of natural resource scarcity: the role of agricultural technologies*. Washington, DC, IFPRI.

Tissier, J. and Grosclaude, J.Y. (2015) Que faut-il penser de l'agriculture climato-intelligente? In: Torquebiau, E. (ed.) *Changement climatique et agricultures du Monde*. Quae, Paris, France, pp. 291–302.

Vanderlinden, J.P. (2015) Adaptation au changement climatique, vulnérabilité et réduction des inégalités. *Revue VRS* 402, 41–42.

World Bank, L. (2008) *Adaptation to Climate Change and Mitigation of Its Impact on Agriculture*. De Boeck Supérieur, Brussels, Belgium, pp. 241–243.

Zambia (2015) Zambia's Intended Nationally Determined Contribution (INDC) to the 2015 agreement on climate change, 12 pp.

General Conclusion

Hubert Cochet*, Olivier Ducourtieux, and Nadège Garambois
*Comparative Agriculture Research & Training Unit, UMR Prodig,
AgroParisTech, Palaiseau, France*

Based on a detailed study of a small number of highly contrasting situations, our work has made it possible to highlight the processes and trajectories that explain the high exposure of certain groups of farmers to hazards, as well as the maintenance or increase of this exposure. This work helps explain the reasons for vulnerability on the agrarian and production system level by revealing the conditions limiting access to resources and means of production, as well as farms' operating methods and outcomes. It highlights and illustrates the influence of past and current agricultural, environmental, and commercial policy choices on the aggravation of vulnerability, and more rarely, on its reduction. Finally, it underlines the adjustment methods and past and current transformations of agricultural and livestock practices that work toward reducing hazard exposure, mitigating vulnerability, and better adapting to the global changes farmers face.

Farmers have long been adapting to very difficult or extremely random climatic conditions for both rainfed and flooded crops (rainfall, rainy season duration, flooding magnitude). In all of the regions studied, farmers have been able to change their practices and rethink, sometimes in depth, the functioning of their agricultural production systems in order to cope with climatic disruptions and more frequent hazards. To do this, today as in the past and in a variety of geographical contexts, they have relied on a detailed knowledge of the ecosystems they use and the expertise they have adapted to soil and climate conditions, resources available, and the economic and social context. The experiences analyzed in this book reveal common levers for adaptation and increasing resilience. They are often based on an agroecological approach: complementary and differentiated use of the different ecosystems to which farmers have access (sometimes at the cost of small-scale development), strengthening agrobiodiversity (varietal diversity, crop associations, diversified cropping patterns, mixed livestock farming), increasing the role of legumes in rainfed agriculture, strengthening links between crops and livestock, and so on.

While these antirisk techniques in the different regions studied mainly concern adaptation to climate change or hazards, they are also at the heart of approaches aimed at better coping with other global changes: demographic growth, increased competition for access to resources, growing integration into market exchanges, changes in relative prices and market fluctuations, deregulation and a decline in public support, and so on. With extremely limited access to equipment and inputs, a situation unfortunately

*Corresponding author: hubert.cochet@agroparistech.fr

©2024 CAB International. *Agrarian Systems and Climate Change: Journeys of adaptation in the Global South* (eds H. Cochet, O. Ducourtieux, and N. Garambois)
DOI: 10.1079/9781800628137.0009

shared by most, it is often thanks to work intensification that the poorest farmers, in spite of everything, have managed to slow down the decline in production and income. Others have succeeded in maintaining and sometimes increasing these, despite the downward trend in the area farmed per worker. Production has been diversified whenever possible, often favoring cash crops, for which these family farms still have some comparative advantages (through market gardening, for example), provided they are not too risky.

In addition to climate change and increasing hazards, since the end of the 1980s, these forms of agriculture have been more directly confronted with competition from world markets than during the post-independence period, while receiving much less support from public authorities, due to a lack of appropriate policy choices and budgetary resources. They are also more exposed to the risks of land grabbing and the development of directly competing forms of agriculture in their immediate vicinity that have incomparably higher levels of labor productivity.

In most of the situations studied, access to sufficient land and irrigation water proved to be a determining factor. This is particularly true when it enables a given family to have access to a diversified range of environmental conditions, making it possible to increase the number of species and varieties cultivated, as well as crop associations and successions, in order to make the best use of the farm labor force, particularly the family. Access to inputs (seeds, synthetic fertilizers, sometimes pesticides) and equipment also makes it possible to increase physical labor productivity during peak workloads: tractors for tilling the soil, power tillers and small combine harvesters, draft power and related equipment for sowing and weeding, carts, small irrigation equipment, storage and post-harvest processing equipment, and so on. This is often a determining factor, as is access to the market under acceptable price conditions.

In view of the still very rapid population growth in most of the regions studied, this progress has not always been sufficient to meet increasing food needs in rural and urban areas. However, given the extremely difficult conditions in which most rural families were placed, which were often worsened by the global changes mentioned above, farmers in these regions have largely demonstrated their "capacity to adapt"

to change, which makes it possible to dispel the image of immobility and inadequacy that is too often attributed to them, particularly by decision makers.

Unfortunately, the main lines of action proposed today by governments or international institutions in public policy documents regarding climate change adaptation remain fixed on this image. Decision makers follow the paths—and dead ends—of the past, not learning from many repeated failures. Policies are still too often limited to a small number of purely technical and often boilerplate recommendations, despite the declared intent of "climate-smart agriculture." The "adaptation" methods proposed to farmers, when not imposed, are supposed to reduce vulnerability, but, on the contrary, they result extremely often in increased risk. While it should be a question of reducing farmers' vulnerability by limiting the risks inherent to climatic deterioration and increased hazards, it is often the innovation itself—the technical advisory package—that leads to increased vulnerability... Moreover, the historical processes behind weakening agricultural systems are neither identified nor taken into account by these policies, nor are the mechanisms of differentiation that make it possible to explain the real causes of some farmers' vulnerability and the reasons for other groups' greater resilience. Thus, specific actions to promote real adaptation could be foreseen. However, vulnerability is above all a social construct, as we have shown through the diachronic and typological approach developed in each of the regions studied in this book. This is why it seems to us that defining vulnerability reduction policies with the aim of better adapting to climate change means making clear political choices that benefit the most vulnerable groups.

In an attempt to reduce farmers' vulnerability and increase their capacity to adapt in the long term, we must rely on the farmers themselves; that is, on their in-depth knowledge of the ecosystems that make up both their living space and their way of working. There is no question about their capacity to innovate, nor their ability to separate the innovations that increase vulnerability from those that reduce it... We, therefore, believe that the priorities are first to identify, in each region, the dynamics under way and pinpoint the practices and innovations that involve both a strategy for limiting risk and an increased capacity for resilience. It is then important to support them through programs that facilitate access to the resources and means of production that farmers need, and by securing the conditions of access to these means of production, including by intervening, when necessary, in the relative price system. Next, it is necessary to reduce uncertainty wherever possible, especially when public action itself is a source of hazard, uncertainty, and vulnerability. This is particularly the case for the most modest producers, as in land tenure systems, for example, or in the authoritarian popularization of risky technical advisory packages imposed from above.

Reflecting on the different means and methods of implementing real climate change adaptation policies in a context of strong demographic growth, scarcity of certain resources, and marked increase in inequalities (particularly concerning access to resources) while taking the need to reduce agriculture's impact on climate change (mitigation) into account, therefore comes down to a complete rethinking of agricultural and rural development policies. There can be no question of limiting ourselves to a single "adaptation component" in a climate policy, however ambitious it may be. Reconciling adaptation, mitigation, and food security, as proposed by proponents of the climate-smart agriculture initiative, is in itself a laudable objective. However, meeting this challenge will only be possible by resolutely attacking the mechanisms that create or amplify inequalities and lack of access to resources and means of production. It is also necessary to reinvent market regulation tools that allow farmers to make a decent living from their activity and investment. Achieving this is only feasible through promoting agriculture that provides employment, is as diversified as possible, and makes only sparing use of synthetic inputs—in short, by moving on from the development of agriculture based on a small number of technical advisory packages or top-down "adaptation techniques" unsuited to farmers' real-life situations.

To imagine that current transformations of practices, farms, landscapes, and territories could be interpreted as solely due to the impact of climate change is, of course, an illusion. Therefore, is the issue of adaptation to climate change still relevant? Is it not a widely instrumentalized smokescreen used to avoid an in-depth review of current agricultural policies and their effects, and justify dressing "business as usual" up in new clothes? By reducing the scope of diagnosis and the proposed solutions to a long list of technical recommendations, the issue of adaptation would thus escape any political questioning.

Index

Note: The page numbers in italics and bold represents figures and tables respectively.